Springer Tracts in Additive Manufacturing

Series Editor

Henrique Almeida, Polytechnik Institute of Leiria, Marinha Grande, Portugal

The book series aims to recognise the innovative nature of additive manufacturing and all its related processes and materials and applications to present current and future developments. The book series will cover a wide scope, comprising new technologies, processes, methods, materials, hardware and software systems, and applications within the field of additive manufacturing and related topics ranging from data processing (design tools, data formats, numerical simulations), materials and multi-materials, new processes or combination of processes, new testing methods for AM parts, process monitoring, standardization, combination of digital and physical fabrication technologies and direct digital fabrication.

More information about this series at http://www.springer.com/series/16694

Shuo Yin · Rocco Lupoi

Cold Spray Additive Manufacturing

From Fundamentals to Applications

Springer

Shuo Yin
Department of Mechanical, Manufacturing and Biomedical Engineering
School of Engineering
Trinity College Dublin
The University of Dublin
Dublin, Ireland

Rocco Lupoi
Department of Mechanical, Manufacturing and Biomedical Engineering
School of Engineering
Trinity College Dublin
The University of Dublin
Dublin, Ireland

ISSN 2730-9576 ISSN 2730-9584 (electronic)
Springer Tracts in Additive Manufacturing
ISBN 978-3-030-73366-7 ISBN 978-3-030-73367-4 (eBook)
https://doi.org/10.1007/978-3-030-73367-4

This Springer imprint is published by the registered company Springer Nature Switzerland AG
The registered company address is: Gewerbestrasse 11, 6330 Cham, Switzerland

Preface

Cold spray is a solid-state material deposition process. It was developed in the 1980s as a coating technology and has been used in a broad range of industries, including aerospace, automotive, energy, medical, marine and other important fields for protecting the underlying substrate from high temperature, corrosion, erosion, oxidation and chemicals. In recent years, time has come to push the boundary of cold spray from coating technology to additive manufacturing. As compared with fusion-based additive manufacturing processes such as LBM, EBM and LMD, cold spray additive manufacturing (CSAM) fabricates deposits at much lower processing temperature. The formation of a cold-sprayed deposit relies mainly on the particle kinetic energy prior to impact rather than the thermal energy. Deposition is achieved through local metallurgical bonding and mechanical interlocking which are caused by localized plastic deformation at the interparticle and particle-substrate interfaces. This consolidation mechanism renders CSAM a unique microstructure that is free of defects commonly encountered in high-temperature processes, such as high residual thermal stress and phase transformation. CSAM offers short production times, unlimited product size and high flexibility. CSAM has been primarily used for fabricating components with simple geometry (e.g., walls, tubes and flanges) and repairing damaged components. The authors have been working in the field of cold spray for more than ten years and thus have a thorough understanding of this novel technology. The aim of this book is to provide a comprehensive introduction of CSAM from its fundamentals to applications based on the authors' research work, some valuable theories and studies form other research groups worldwide. The book covers all critical topics related to CSAM including working principle, fluid dynamics, microstructures, strengthening strategy, manufacturing process, new materials, applications and future perspective. Some of these topics have never been covered in previously published books. This book can be served as a valuable reading material for a broad range of audiences including researchers, business professionals, engineers, undergraduate and graduate students, technical consultants and administrators, general public. A quick preview of the book contents is given as follows:

In Chap. 1, the working principles of CSAM, commercial cold spray system, comparison between CSAM and other additive manufacturing processes and brief introduction of the applications of CSAM are presented.

Chapter 2 provides a comprehensive introduction on the gas flow, particle acceleration, and heat transfer behavior in cold spray. In CSAM, deposition is achieved through the high-velocity impact of powder particles onto substrate, and the acceleration of the particles is released through a convergent–divergent nozzle. Therefore, a thorough understanding of particle acceleration and heating behavior inside and outside a cold spray nozzle is critical for developing high-performance cold-sprayed deposits. This chapter summarizes and reviews the most important scientific work with regard to experimental measurements and numerical simulations in the topic of aerodynamics that are involved in cold spray.

In Chap. 3, key manufacturing parameters that are critical for the operation of CSAM are introduced, including gas pressure, gas temperature, gas type, powder feed rate, nozzle traverse speed, scanning step, standoff distance, spray angle and trajectory. Their roles in cold spray deposition are introduced and the relationship among these parameters are discussed. These parameters must work in unison and under careful control to produce high-quality deposits with desirable microstructure and excellent performance.

In Chap. 4, the microstructures of cold-sprayed deposits are interpreted and discussed. Cold-sprayed deposits typically have heterogenous microstructure and grain structures. A large number of dislocations and newly formed grains due to the severe plastic deformation are localized in the shear zones at the interparticle interfaces, while in the interior region the microstructure is mainly characterized by coarse grains. This chapter thoroughly discusses the formation mechanism of dislocations and ultra fine grains (UFGs) in various cold-sprayed deposits. The formation mechanisms of intermetallic compounds and amorphous phases are also discussed.

Chapter 5 provides an introduction of commonly used strengthening strategies for improving the mechanical properties of cold sprayed deposits. Cold sprayed deposits normally have unfavorable mechanical properties in their as-fabricated state as compared to conventional fusion-based additive manufactured counterparts due to the inherent defects (i.e., porosity and unbonded interparticle boundary) in the cold sprayed deposits. As structural weaknesses, theses defects will impair the mechanical properties of cold sprayed deposit. In this chapter, various strengthening methods including conventional annealing, hot isostatic pressing, hot rolling, in-situ shot peening, in-situ densification, in-situ laser assistance, friction stirring and powder heat treatment are introduced and their strengthening mechanisms are discussed.

Chapters 6–8 aim to introduce the application of CSAM for fabricating special materials including metal matrix composites (MMCs), nanocrystalline metals, amorphous metals and high entropy alloys (HEAs). These materials are either composite materials or non-conventional metals and alloys; they are mostly difficult to be deposited via cold spray due to the nature of the starting powder feedstock. For example, the feedstock for fabricating MMC deposits must be a mixture of two or more dissimilar materials, and it is hard to accurately control the ratio between different components in the deposit. For nanocrystalline metals, amorphous metals and HEAs, the starting powders typically have high hardness and strength, which significantly reduces their cold-sprayability and potentially leads to the deposit having a large number of defects. These chapters comprehensively explain how these unconventional metallic materials are fabricated via cold spray and their microstructure and properties after cold spray.

In Chap. 9, the current and potential applications of CSAM in industry are presented. CSAM can be used in a broad range of industrial sectors such as aerospace, aviation, marine, automotive and nuclear. It is capable of fabricating free-stranding components and repairing damage components. Cold-sprayed deposits typically have a rough, undulating and porous surface and low dimensional accuracy, and thus they are not suitable for immediate use. This necessitates the use of a machining or finishing process. Therefore, this chapter aims to provide a thorough discussion on the post machining and the industrial applications of cold sprayed deposits.

Chapter 10 summarizes the main findings and contributions of this book and also provides some important suggestions and outlooks for the future development of CSAM.

At the end, the authors gratefully appreciate your interests in reading this book and hope the book can provide you a better understanding of the newly developed CSAM technology.

Dublin, Ireland
January 2021

Dr. Shuo Yin
Dr. Rocco Lupoi

Contents

About the Authors

Dr. Shuo Yin was appointed Assistant Professor and Principle Investigator within the Department of Mechanical, Manufacturing and Biomedical Engineering in January 2019. His current research interests mainly focus on the experimental and numerical studies of additive manufacturing processes including cold spraying, selective laser melting, digital light processing and direct-ink writing. He received B.Eng. and Ph.D. from Dalian University of Technology in July 2007 and January 2013. From 2013 to 2015, he worked as a Postdoctoral Researcher at University of Technology of Belfort-Montebiard in France and Wageningen University in the Netherland. He then moved to Dublin and was appointed Research Fellow in Trinity College Dublin. He is an awardee the Government of Ireland Postdoctoral (GOIPD) Research Fellowship. Dr. Yin's research has been financially supported by a number of funding bodies including Irish Research Council, Science Foundation Ireland, Enterprise Ireland, InterTradeIreland and industrial partners. He has authored or co-authored over 100 papers in peer-reviewed international journals including *Progress in Materials Science, Acta Materialia, Scripta Materialia, Additive Manufacturing, Materials & Design, Composites Part B: Engineering*, etc., and four chapters. Among these publications, three of them were highlighted by the journals (*Materials and Design, Journal of Thermal Spray Technology*) as cover pages. His publications have been cited by more than 2600 times, including a hot paper (top 0.1%) and two highly cited papers (top 1%). He has been serving as guest editors for *Journal of Thermal Spray Technology* and *Coatings* and reviewer of more than 30 journals.

Dr. Rocco Lupoi graduated as mechanical engineer in 2004 from Polytechnic University of Turin and obtained Ph.D. with great success from the University of Bath (UK) in 2008. His Ph.D. project was covered by an article with dedicated front cover in EUREKA magazine (May 2005) with title devices that absorb energy on demand. He really enjoyed research and formed a clear passion for the development of new manufacturing processes. As such, he was hired as postdoctoral researcher in Cambridge University (UK) in 2008 to carry out ground-breaking work in a new field, now known as additive manufacturing. In Cambridge, he published key papers, submitted a patent (now granted) and formed a spin-off company with co-workers (Laser Fusion Tech.) to commercialize the processes he had developed. Despite the commercial attraction, academic research remained his true passion; in summer 2012, he accepted the post of Assistant Professor in Trinity College Dublin and started to develop his very own group. In 2013, he became Marie Curie Fellow (CIG scheme), and in 2014, he was awarded of a key grant from the European Space Agency (Technology Research Programme—TRP). He was the very first PI in Trinity to have secured an award of this size from ESA. The project was featured in the national innovation news page of the Irish Times, January 19, 2015. As an independent PI, he is now a Funded Investigator in AMBER (the national research centre in advanced materials and bioengineering) and I-Form (the national centre in advanced manufacturing) and has published >120 peer-reviewed papers. He is now an Associate Member of the International Academy of Production Engineering (CIRP). In recognition to his research career, Dr. Lupoi was elected Fellow of Trinity College Dublin in 2019.

Chapter 1
Introduction to Cold Spray Additive Manufacturing

Abstract Cold spray is a solid-state coating deposition technology which has recently been applied as an additive manufacturing process to fabricate individual components and to repair damaged components. In comparison with fusion-based high-temperature additive manufacturing processes, cold spray additive manufacturing (CSAM) has shown to retain the original properties of the feedstock, to produce oxide-free deposits, and to not adversely influence underlying substrate materials during manufacture. Therefore, CSAM is attracting considerable attention from both scientific and industrial communities. Although CSAM is an emerging additive manufacturing technology, a body of work has been carried out by the authors' and other research groups worldwide, and the technology has been applied across a range of manufacturing areas. This chapter aims to provide a brief introduction of CSAM and acts as a start of this book.

1.1 Cold Spray Process Principle

Cold spray is a solid-state material deposition process, which was originally developed as a coating technology in the 1980s [1, 2]. In this process, high-temperature compressed gases (typically nitrogen, air, or helium) are used as the propulsive gas to accelerate powder feedstock (typically metals and metal matrix composites) to a high velocity (typically higher than 300 m/s), and to induce deposition when the powders impact onto a substrate (typically metals). In contrast to conventional high-temperature deposition processes, the formation of a cold spray deposit relies largely on the particle kinetic energy prior to impact rather than the thermal energy. The feedstock used for cold spray remains solid state during the entire deposition process. Deposition is achieved through local metallurgical bonding and mechanical interlocking which are caused by localized plastic deformation at the inter-particle and particle–substrate interfaces. This allows for the avoidance of defects commonly encountered in high-temperature deposition processes, such as oxidation, residual thermal stress and phase transformation [3–5].

Successful deposition of a cold sprayed deposit requires the feedstock particles to exceed a critical impact velocity [6–13]. In cold spray, the formation of a deposit

S. Yin and R. Lupoi, *Cold Spray Additive Manufacturing*, Springer Tracts in Additive Manufacturing, https://doi.org/10.1007/978-3-030-73367-4_1

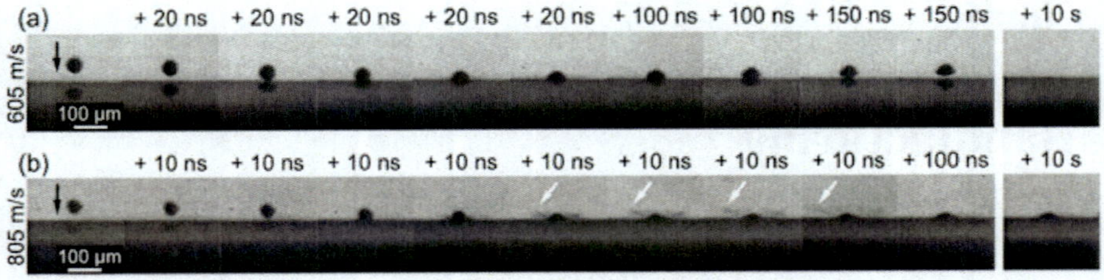

Fig. 1.1 In-situ observation of a 45 μm aluminium particle impacting onto an aluminium substrate at the velocity of (a) 605 m/s and (b) 805 m/s. The critical velocity is between 605 and 805 m/s. [6]

consists of two different stages. The first stage involves the deposition of an initial layer of particles where bonding occurs between feedstock particles and the substrate material; the second stage is the deposition on top of the layer(s) previously deposited, where bonding occurs between feedstock particles. Each stage has a respective critical velocity, i.e. for particle/substrate bonding and for deposit growth. Particle impact velocity must satisfy both criterions for successful deposition. In general, a higher particle velocity will result in improved deposit quality. Note that when the powder and substrate are the same material, the critical velocity can be considered as the same for both stages. Figure 1.1 shows an in-situ observation of a 45 μm aluminium particle impacting onto an aluminium substrate at different velocities, where particle deposition only occurs when the impact velocity is beyond the critical velocity [6]. Critical velocity is not a constant but depends on several factors including material type, particle size, and particle temperature. Larger particles or higher particle temperature upon impact helps to reduce critical velocity [6, 10, 12–15].

Feedstock for cold spray is typically gas-atomized spherical metal powders, and in some cases irregular powders [5]. The powder size must fall into a specific range, normally between 10 and 100 μm in diameter, to guarantee a sufficiently high particle impact velocity that is beyond the critical velocity [11]. Particles with diameters greater than 100 μm or lower than 10 μm are difficult to accelerate with the propulsive gas and thus usually fail to deposit [16]. For most metals, the most frequently used size-range is 20 - 60 μm; for low-density metals such as aluminium and zinc, the upper limit of the size-range can reach 100 μm [11].

1.2 Cold Spray System

The cold spray process can be divided into two categories, according to the pressure of the propulsive gas: high pressure cold spray (>1 MPa), and low pressure cold spray (<1 MPa). Figure 1.2a shows a schematic of a high pressure cold spray system. Compressed gas is divided into two streams upon entering the cold spray system. One stream of the compressed gas (referred to as the propulsive gas) passes through a gas heater, where it is heated to a high temperature, At the same time, the second stream of the compressed gas (referred to as the carrier gas) passes through a powder

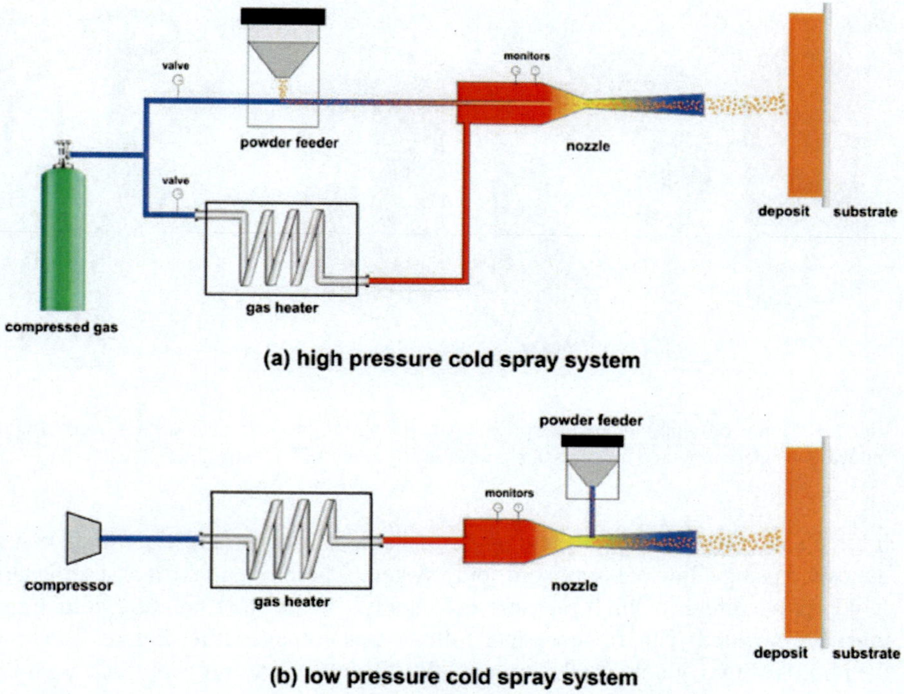

Fig. 1.2 Schematic of high pressure and low pressure cold spray systems

feeder, where it becomes laden with the feedstock particles. These two gas streams are then mixed, before entering a de-Laval nozzle, where the gas expands to generate a supersonic gas and powder stream. In order to make certain there is successful injection of the powder into the mixed gas stream the carrier gas pressure must be slightly higher than the propulsive gas pressure. These high-velocity particles impact onto a substrate to form a coating or thick deposit at a temperature well below their melting temperature. Generally, high pressure cold spray systems can operate at the same pressures as low pressure cold spray systems, but the reverse is not true due to the special design of low pressure cold spray systems. Figure 1.2b shows a schematic of a low pressure cold spray system. Two obvious differences in the low pressure system from the high pressure system are: the high-pressure compressed gas is normally replaced by a portable air compressor; the powder injection point is at the nozzle divergent section where the local gas pressure is sufficiently low to allow the release of powders from the powder feeder at the atmospheric pressure. These differences make low pressure cold spray systems more flexible and much cheaper in both equipment and processing costs than a high pressure cold spray system [3, 5]. They are particularly suitable for the restoration of damaged components due to the portability of the simpler low pressure cold spray systems [17, 18]. However, as the particle velocity in low pressure cold spray systems is much lower than in high pressure cold spray systems, the application of low pressure cold spray is significantly

Fig. 1.3 Photos of commercial cold spray apparatus. **a** high pressure cold spray system (Impact Innovations, Germany) and low pressure cold spray system (SST, CentreLine, Canada)

limited and only a limited range of materials such as copper and aluminium can be deposited using a low pressure cold spray system. Therefore, in most cases, the term 'cold spray' refers to 'high pressure cold spray', while 'low pressure cold spray' must be specified. This review paper follows this nomenclature. Figure 1.3 shows the photos of two widely used commercial cold spray systems developed by Impact Innovations (Germany) and CenterLine (Canada).

1.3 Cold Spray Additive Manufacturing

Cold spray has been widely applied as a coating technology in a broad range of industries, including aerospace, automotive, energy, medical, marine and other fields. Cold sprayed deposits provide effective protection against high temperature, corrosion, erosion, oxidation and chemicals [19–21]. In recent years, cold spray has been successfully developed as an additive manufacturing technology to fabricate free-standing metal components, and to restore damaged metal components [4, 22, 23]. This development shines a new light on conventional additive manufacturing technologies, and significantly broadens the applications of cold spray.

As a new member of the additive manufacturing family, CSAM has all the benefits of cold spray technology. In comparison with other well-investigated fusion-based additive manufacturing technologies such as laser beam melting (LBM), electron beam melting (EBM) and laser metal deposition (LMD) [24, 25], CSAM has unique advantages. Table 1.1 compares the advantages and disadvantages of CSAM and other existing additive manufacturing technologies. As can be seen, the most important benefits of CSAM over other additive manufacturing processes are shorter production times, unlimited product size, high flexibility, and suitability for damaged component repair. Furthermore, CSAM is particularly suitable for high-reflectivity metals such as copper and aluminium, which are very challenging to manufacture

Table 1.1 Comparison of advantages and disadvantages between CSAM and other fusion-based additive manufacturing

	CSAM	SLM	EBM	LMD
Powder feed mode	Direct deposition	Powder bed	Powder bed	Direct deposition
Feedstock limitations	Difficulty processing high hardness and strength metals	Difficulty processing high reflectivity and poor flowability metals	Unsuitable for non-conductive and low melting-temperature metals	Difficulty with high reflectivity metals
Powder melting	No	Yes	Yes	Yes
Product size	Large	Limited	Limited	Large
Dimensional accuracy	Low	High	High	Medium
Mechanical properties (AF)	Low	High	High	High
Mechanical properties (HT)	High	High	High	High
Production time	Short	Long	Long	Long
Equipment flexibility	High	Low	Low	Low
Suitable for repair	Yes	No	No	Yes

AF as-fabricated, *HT* heat treatment

using laser based additive manufacturing processes. However, the disadvantages of CSAM are also obvious. CSAM typically fabricates a semi-finished product with a rough surface, which requires post-machining. In addition, CSAM deposits possess poorer mechanical properties in their as-fabricated state due to the inherent defects. As a result, heat treatments are commonly applied to improve the mechanical properties.

Currently, CSAM has been primarily used for fabricating components with rotational symmetry such as cylinder walls, tubes and flanges. With the aid of a well-designed mandrel, spindle, or mask, it is also applicable for the fabrication of complex structures such as array fin heat exchangers, metal labels and metal 2D codes [19]. CSAM is also widely used to repair damaged components due to its high flexibility, lack of adverse effects on the underlying substrate material, and to retain the original properties of the feedstock. So far, CSAM has been successfully applied to restore a variety of corroded and mechanically damaged components in various fields, particularly in the aerospace industry. Repair using CSAM returns the components to serviceable condition, significantly reducing the cost when compared to replacement with a new component [26].

References

1. Alkhimov, A.P., Kosarev, V.F., Nesterovich, N.I., Papyrin., A.N.: Method of applying coatings (SU 1618778) (1980)
2. Alkhimov, A.P., Kosareve, V.F., Papyrin, A.N.: A method of cold gas-dynamic spray deposition. Dokl. Akad. Nauk SSSR. **315**, 1062–1065 (1990)
3. Assadi, H., Kreye, H., Gärtner, F., Klassen, T.: Cold spraying—a materials perspective. Acta Mater. **116**, 382–407 (2016). https://doi.org/10.1016/j.actamat.2016.06.034
4. Raoelison, R.N., Xie, Y., Sapanathan, T., Planche, M.P., Kromer, R., Costil, S., Langlade, C.: Cold gas dynamic spray technology: a comprehensive review of processing conditions for various technological developments till to date. Addit. Manuf. **19**, 134–159 (2018)
5. Rokni, M.R., Nutt, S.R., Widener, C.A., Champagne, V.K., Hrabe, R.H.: Review of relationship between particle deformation, coating microstructure, and properties in high-pressure cold spray. J. Therm. Spray Technol. 1–48 (2017). https://doi.org/https://doi.org/10.1007/s11666-017-0575-0
6. Hassani-Gangaraj, M., Veysset, D., Nelson, K.A., Schuh, C.: In-situ observations of single micro-particle impact bonding. Scr. Mater. **145**, 9–13 (2018). https://doi.org/10.1016/j.scriptamat.2017.09.042
7. Meng, F., Yue, S., Song, J.: Quantitative prediction of critical velocity and deposition efficiency in cold-spray: a finite-element study. Scr. Mater. **107**, 83–87 (2015). https://doi.org/10.1016/j.scriptamat.2015.05.026
8. Assadi, H., Gärtner, F., Stoltenhoff, T., Kreye, H.: Bonding mechanism in cold gas spraying. Acta Mater. **51**, 4379–4394 (2003). https://doi.org/10.1016/S1359-6454(03)00274-X
9. Raletz, F.F., Vardelle, M., Ezo'o, G.: Critical particle velocity under cold spray conditions. Surf. Coatings Technol. **201**, 1942–1947 (2006). https://doi.org/10.1016/j.surfcoat.2006.04.061
10. Hassani-Gangaraj, M., Veysset, D., Nelson, K.A., Schuh, C.A.: Melting can hinder impact-induced adhesion. Phys. Rev. Lett. **119**, 175701 (2017). https://doi.org/10.1103/PhysRevLett.119.175701
11. Grujicic, M., Zhao, C.L.C., DeRosset, W.S.W., Helfritch, D.: Adiabatic shear instability based mechanism for particles/substrate bonding in the cold-gas dynamic-spray process. Mater. Des. **25**, 681–688 (2004). https://doi.org/10.1016/j.matdes.2004.03.008
12. Assadi, H., Schmidt, T., Richter, H., Kliemann, J.O., Binder, K., Gärtner, F., Klassen, T., Kreye, H.: On parameter selection in cold spraying. J. Therm. Spray Technol. **20**, 1161–1176 (2011). https://doi.org/10.1007/s11666-011-9662-9
13. Schmidt, T., Gärtner, F., Assadi, H., Kreye, H.: Development of a generalized parameter window for cold spray deposition. Acta Mater. **54**, 729–742 (2006). https://doi.org/10.1016/j.actamat.2005.10.005
14. Bae, G., Kumar, S., Yoon, S., Kang, K., Na, H., Kim, H.J., Lee, C.: Bonding features and associated mechanisms in kinetic sprayed titanium coatings. Acta Mater. **57**, 5654–5666 (2009). https://doi.org/10.1016/j.actamat.2009.07.061
15. Yin, S., Wang, X., Suo, X., Liao, H., Guo, Z., Li, W., Coddet, C.: Deposition behavior of thermally softened copper particles in cold spraying. Acta Mater. **61**, 5105–5118 (2013). https://doi.org/10.1016/j.actamat.2013.04.041
16. Yin, S., Meyer, M., Li, W., Liao, H., Lupoi, R.: Gas Flow, Particle Acceleration, and Heat Transfer in Cold Spray: A review. J. Therm. Spray Technol. **25**, 1–23 (2016). https://doi.org/10.1007/s11666-016-0406-8
17. Villafuerte, J.: Current and future applications of cold spray technology. Met. Finish. **108**, 37–39 (2010)

18. Kashirin, A., Klyuev, O., Buzdygar, T., Shkodkin, A.: Modern applications of the low pressure cold spray. In: Proceedings of the 2011 International Thermal Spray Conference. , Hamburg (2011)
19. Champagne, V., Helfritch, D.: The unique abilities of cold spray deposition. Int. Mater. Rev. **6608**, 1–19 (2016). https://doi.org/10.1080/09506608.2016.1194948
20. Vilardell, A.M., Cinca, N., Concustell, A., Dosta, S., Cano, I.G., Guilemany, J.M.: Cold spray as an emerging technology for biocompatible and antibacterial coatings: state of art. J. Mater. Sci. 4441–4462 (2015). https://doi.org/https://doi.org/10.1007/s10853-015-9013-1
21. Champagne, V.K., Helfritch, D.J.: Mainstreaming cold spray—push for applications. Surf. Eng. **30**, 396–403 (2014). https://doi.org/10.1179/1743294414Y.0000000277
22. Li, W., Yang, K., Yin, S., Yang, X., Xu, Y., Lupoi, R.: Solid-state additive manufacturing and repairing by cold spraying: a review. J. Mater. Sci. Technol. **34**, 440–457 (2018). https://doi.org/10.1016/j.jmst.2017.09.015
23. Yin, S., Aldwell, B., Lupoi, R.: Cold spray additive manufacture and component restoration. 195–224 (2018)
24. Herzog, D., Seyda, V., Wycisk, E., Emmelmann, C.: Additive manufacturing of metals. Acta Mater. **117**, 371–392 (2016). https://doi.org/10.1016/j.actamat.2016.07.019
25. Debroy, T., Wei, H.L., Zuback, J.S., Mukherjee, T., Elmer, J.W., Milewski, J.O., Beese, A.M., Wilson-heid, A., De, A., Zhang, W.: Additive manufacturing of metallic components—process, structure and properties. Prog. Mater. Sci. **92**, 112–224 (2017). https://doi.org/10.1016/J.PMATSCI.2017.10.001
26. Lead, T., Boston, C.: Cold spray metallization technology focus area status fred lancaster NAVAIR cold spray (2017)

Chapter 2
Gas Flow, Particle Acceleration and Heat Transfer in Cold Spray Additive Manufacturing

Abstract In CSAM, successful particle deposition requires the particles to achieve a higher velocity than critical velocity. Therefore, a thorough understanding of particle acceleration and heating behavior inside and outside a cold spray nozzle is critical for developing high-performance cold sprayed deposits. Over the years, much effort has been devoted to investigate the supersonic gas flow and the consequent particle acceleration and heating behavior inside and outside the nozzle. Experimental investigation is the most direct way to clarify these physical phenomena involved in cold spray process. However, the relatively high money- and timing- cost, especially the infeasibility to capture all the flow features (e.g. flow velocity, temperature, density and turbulence properties) inside and outside the nozzle significantly limits the wide application of experimental approach. To deal with this problem, analytical and numerical modeling were developed and employed in many studies. In the early stage, various analytical models were developed to predict the gas flow properties and particle velocity. These analytical models, mainly one-dimensional (1D) model, normally introduced assumptions and simplifications to the real physical problems, thus the prediction accuracy may be not satisfactory. Thanks to the rapid growth of the computer power, computational fluid dynamics (CFD) technique has become a popular approach to predict the gas flow properties and particle velocity in cold spray. It is highly flexible to simulate the gas flow at different working conditions and costs less than experiment. This chapter provides a comprehensive introduction on the gas flow, particle acceleration, and heat transfer behavior in cold spray.

2.1 Analytical Models

2.1.1 Gas Flow Models

Analytical models based on 1D isentropic assumption were widely used for the prediction of the gas flow features inside the cold spray nozzle in the early stage and are still being used today [1–16]. For a specific nozzle with fixed expansion ratio, the Mach number at any position along the nozzle axis can be obtained by the following equations,

S. Yin and R. Lupoi, *Cold Spray Additive Manufacturing*, Springer Tracts in Additive Manufacturing, https://doi.org/10.1007/978-3-030-73367-4_2

$$\frac{A}{A^*} = \frac{1}{M}\left[\left(\frac{1}{\gamma+1}\right)\left(1+\frac{\gamma-1}{2}M^2\right)\right]^{\frac{\gamma+1}{2(\gamma-1)}} \tag{2.1}$$

where M, A, A*, A/A* and γ are the Mach number, nozzle cross-sectional area, nozzle cross-sectional area at the throat, nozzle expansion ratio and ratio of gas specific heats. Once the Mach number is obtained, the corresponding state variable of the gas inside the nozzle can be derived from the following isentropic equations,

$$\frac{P_0}{P} = \left(1+\frac{\gamma-1}{2}M^2\right)^{\frac{\gamma}{\gamma-1}} \tag{2.2}$$

$$\frac{T_0}{T} = 1+\frac{\gamma-1}{2}M^2 \tag{2.3}$$

$$\frac{\rho_0}{\rho} = \left(1+\frac{\gamma-1}{2}M^2\right)^{\frac{1}{\gamma-1}} \tag{2.4}$$

$$v = M\sqrt{\gamma RT} \tag{2.5}$$

where P_0, P, T_0, T, ρ_0, ρ, v and R are the gas stagnation pressure, gas pressure, gas stagnation temperature, gas temperature, gas stagnation density, gas density, gas velocity and gas constant. Having the working conditions (P_0 and T_0) and nozzle expansion ratio, the detailed flow properties along the nozzle axis can be calculated by Eqs. 2.1–2.5. The 1D isentropic model is simple and results in a prediction with tolerable engineering error, thus attracting many users. However, a significant shortcoming of the model is that it does not take the ambient pressure into consideration [2, 5, 6]. A comparison work shown in Fig. 2.1 indicates that the 1D isentropic model led to large errors, particularly when the stagnation pressure was low [9]. the model also fails to include the viscosity, turbulence, boundary layer, nozzle length, outside supersonic jet and the compressed layer (bow shock) on the substrate surface, which

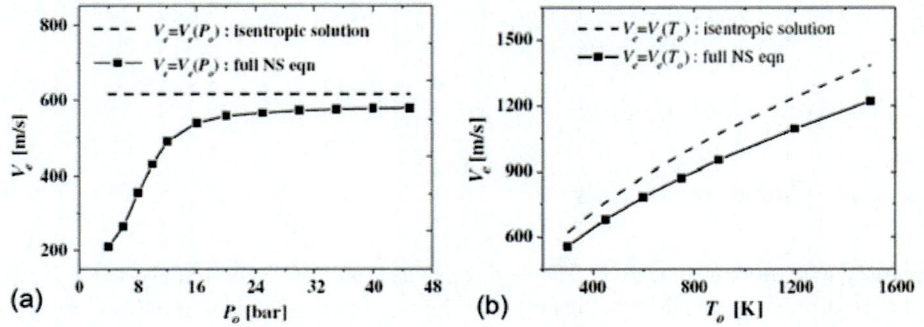

Fig. 2.1 Comparison of the gas pressure and velocity at the nozzle exit against stagnation pressure and temperature between 1D isentropic analysis and CFD modeling with the same working conditions

limits its wider application in the scientific field. A more realistic model considering the presence of boundary layers along the nozzle wall and bow shock on the substrate surface was developed by Alkhimov et al. [17]. This model further improved the 1D isentropic model, but complicated equations and additional experimental data required as the input of the computation significantly limits its wide application. Similarly, Kosarev et al. also developed an analytical model to describe the gas flow properties in cold spray, where two-dimensional (2D) velocity distribution at the cross-section, the outside supersonic jet and compressed bow shock were represented by the pure mathematical equations or empirical equations [18]. Good agreement between the analytical results and experiment measurements was achieved. Despite this, it is still not quite applicable to the general people or engineer who do not have strong background in mathematic.

In summary, 1D isentropic model is a convenient tool to roughly estimate the gas flow properties inside the nozzle, but the prediction accuracy cannot be completely guaranteed due to the effect of ambient pressure, also the nozzle outside supersonic flow and bow shock are not included. More realistic model may result in better estimation of the nozzle inside and outside flow, but such models are normally developed based on complicated equations that are hard to understand, significantly limiting their applications by the general users. Therefore, it is still recommended to use 1D isentropic model for roughly estimating nozzle inside flow under the condition that the ambient pressure has no effect. For a better prediction accuracy, the solution of full Navier–Stokes (NS) equations by CFD technique may be the best choice.

2.1.2 *Particle Motion Models*

Particulate phase in cold spray is normally assumed sufficiently dilute and moves along the nozzle axis. In this case, the particle velocity can be calculated by the following equation,

$$m_p \frac{dv_p}{dt} = C_D \rho (v - v_p) |v - v_p| \frac{A_p}{2} \tag{2.6}$$

where C_D, m_p, v_p, t and A_p, are the drag coefficient, particle mass, particle velocity, time and cross-sectional area of particle. Note that drag coefficient is an important parameter affecting particle motion in the gas flow that will be detailed discussed in the following text. Here, if the gas velocity, gas density and drag coefficient are assumed to be constant, the following simplified expression for roughly estimating the particle velocity can be obtained [1, 2, 6],

$$\log\left(\frac{v - v_p}{v}\right) + \frac{v}{v - v_p} - 1 = \frac{C_D A_P \rho l}{2m_p} \tag{2.7}$$

where l is the axial position from the nozzle inlet. Equation 2.7 combined with Eqs. 2.1 and 2.5 clearly reveals that particle velocity inside the nozzle is affected by particle size, particle density, gas stagnation temperature, gas molecular weight and nozzle shape [6], which have been confirmed to be correct by CFD modeling and experimental validation. However, Eq. 2.7 was proposed on the basis of many assumptions and hence may generally not be valid for accurate perdition and was not widely used so far. Besides, Eqs. 2.1–2.5 are only available for the nozzle inside flow, thus these equations (Eqs. 2.1, 2.5 and 2.7) can only be used to calculate the particle velocity inside the nozzle. In reality, after particles leave the nozzle exit, they will pass a series of shockwaves and a bow shock in front of the substrate; their velocities upon the impact on the substrate are not the same as at the nozzle. In order to deal with the effect of the nozzle outside flow on the particle motion, an empirical model is proposed based on experimental and computational analyses [2]. The equations are given as following,

$$v_p^{impact} = v_p^{exit} e^{-3\rho^{st} L^{st}/4\rho_p d_p} \tag{2.8}$$

$$\rho^{st} = R_e\left(-1.04 + 2.27M_e - 0.21M_e^2\right) \tag{2.9}$$

$$L^{st} = R_e(0.97 - 0.02M_e) \quad 1 \le M_e \le 5 \tag{2.10}$$

where ρ^{st}, L^{st}, v_p^{exit}, v_p^{impact}, Re and M_e are the average gas density in the stagnant region, thickness of the stagnation region, particle velocity at nozzle exit, particle velocity upon impact, gas Reynold number and Mach number at the nozzle exit. This model allows estimating the particle velocity upon impact through the particle properties at the nozzle exit, but its accuracy was not widely validated. Similarity, Assadi et al. also proposed a series of empirical equations to estimate the gas velocity and particle velocity by fitting the numerical calculated data with parametric expressions [19],

$$v_g = \left(c_1 + \frac{d^* - c_1 d^*}{d_e}\right)^e c^* \tag{2.11}$$

$$v_p^{exit} = \left(\frac{c_2}{\sqrt{RT_0}} + \sqrt{\frac{\rho_p d_p}{C_D L_d P_0}}\right)^{-1} \tag{2.12}$$

$$v_p^{impact} = v_p^{exit}\left(1 + \frac{\rho_0 \delta}{\rho_{pd_p}}\right)^{-1} \tag{2.13}$$

where d^*, d_e, c^* and L_d are the nozzle throat diameter, nozzle exit diameter, sound speed at the nozzle throat and nozzle divergent length, respectively. c_1, c_2, e and δ are the fitting parameters, which can only be obtained from the aforehand numerical calculation. Although the predicted results from this model showed good agreement with experiment [19], the determination of those fitting parameters appearing in the expression lowers its feasibility.

In a word, analytical models for particle velocity prediction are convenient in comparison with numerical modeling and experimental measurement. However, many assumptions on the isentropic model make the prediction results unreliable. Empirical equations are able to provide more realistic prediction and capable to consider the effect of the nozzle outside flow on particle velocity, but they are still not quite feasible because of the aforehand determination of fitting parameters. In addition, all of the existing analytical models for predicting particle velocity are based on 1D assumption, which means the particle distribution at the cross-section cannot be considered. As the computer's power growing fast today, numerical solution of the 2D or 3D full NS equations is becoming faster and faster. Under such circumstances, analytical models do not have too many advantages compared with the CFD modeling in both prediction accuracy and time consumption.

2.2 CDF Modeling

2.2.1 Computational Methodology

Solution of the gas flow

The governing equations to be solved for the gas flow are NS equations, which comprise of the conservation of mass, momentum and energy,

$$\frac{\partial \rho}{\partial t} + \frac{\partial}{\partial x_j}(\rho u_j) = 0 \tag{2.14}$$

$$\frac{\partial}{\partial t}(\rho u_i) + \frac{\partial}{\partial x_j}(\rho u_i u_j) = \frac{\partial}{\partial x_j}\left[-p\delta_{ij} + \mu\left(\frac{\partial u_i}{\partial x_j} + \frac{\partial u_j}{\partial x_i}\right)\right] + S_M \tag{2.15}$$

$$\frac{\partial}{\partial t}(\rho C_P T) + \frac{\partial}{\partial x_j}(\rho u_j C_P T) = \frac{\partial}{\partial x_j}\left(\lambda \frac{\partial T}{\partial x_j}\right) + S_T \tag{2.16}$$

where $t, \rho, u_i, P, \mu, T, C_p, S_M, S_T$ are the time, gas density, gas velocity components, gas static pressure, dynamic viscosity, gas temperature, specific heat at constant pressure and additional source terms in momentum and energy, respectively [20]. In cold spray modeling, the flow inside and outside the nozzle was normally regarded as steady-state. Because the gas speed is supersonic, the equation of state for compressible flow is required to close the NS equation set, which was generally taken for ideal gas law. The source terms appearing in the momentum and energy equations were often neglected. However, under certain conditions, e.g. high powder feed rate, the momentum and energy transfer between the gas and the particulate phase can be important, which can be accounted for by a respective phase coupling through these terms [8, 21–24]. The computational domain was normally divided with structured mesh as it is more efficient and has better convergence capability over the unstructured mesh. In some 3D modeling, unstructured mesh was also employed since it is more suitable in dealing with complicated geometry [8, 25–27]. As for the boundary conditions, Table 2.1 lists the most commonly used boundary conditions and its mathematic description for modeling the impinging jet in cold spraying. A principle for proper boundary is that the surrounding atmosphere boundary should be arranged sufficiently far from the concerned region (free or impinging jet) to eliminate the influence on the computational result. The discretization scheme of the NS equation set is also very important. Basically, low-order schemes result in low accuracy but help the convergence and save the computational time, while high-order schemes give the reverse outcome. So far, it seems no publications had particular attention to the selection of discretization scheme. But the discretization scheme indeed posed impact on the prediction accuracy of supersonic compressible flow, particularly on the shock-wave structure [28]. For the thermophysical parameters, the gas viscosity was considered as a constant in most previous studies, but the temperature-dependent Sutherland law was also used in some works because the variation of viscosity with temperature was very prominent in high-speed compressible cold spraying flows [29–35]. Other thermophysical properties, e.g. specific heat and thermal conductivity, were mostly assumed constants in the simulation because they were not quite sensitive to the temperature and compressibility. While in the works of Zahiri et al., the specific heat and thermal conductivity were considered as temperature-dependent parameters [36, 37].

Turbulence model is one of the most important factors that affects the modeling accuracy. Table 2.2 summarizes the ever-used turbulence models for modeling the gas flow in cold spraying. Most of the turbulence models found in literature were based on the closure of the Reynold Averaged Navier–Stokes (RANS) equations. The Spalart–Allmaras, as a 1-equation model, was initially designed for aerospace applications involving wall-bounded flows, e.g. aerodynamic flows. But it produces relatively large errors for free shear flows, especially the plane and round jet which are the case of cold spray, thus it was barely used so far. The standard k-ε model is the most frequently used model because it is applicable in various flows, relatively simple to implement and normally leads to a stable solution that converges easily. The RNG

Table. 2.1 Boundary conditions for modelling the impinging jet in cold spraying

	P	v	T
Nozzle/injector inlet	Specified	$\frac{\partial v}{\partial n} = 0$	Specified
Surrounding atmosphere	Specified	$\frac{\partial v}{\partial n} = 0$	$\frac{\partial T}{\partial n} = 0$
Nozzle walls	$\frac{\partial P}{\partial n} = 0$	0	$\frac{\partial T}{\partial n} = 0$
Substrate	$\frac{\partial P}{\partial n} = 0$	0	$\frac{\partial T}{\partial n} = 0$
Symmetrical axis (for 2D axisymmetric model)	$\frac{\partial P}{\partial r} = 0$	$\frac{\partial v}{\partial r} = 0$	$\frac{\partial T}{\partial r} = 0$

Table. 2.2 Summary of the turbulence models used for modeling the supersonic gas flow in cold spraying

Turbulence models	Model theory	Equations	References
Standard k-ε	RANS	2	[3, 8, 22, 24, 40, 41, 43–62]
RNG k-ε	RANS	2	[9, 11, 14, 26, 62–72]
Realizable k-ε	RANS	2	[10, 25, 27, 30, 41, 73–77]
RSM	RANS	7	[7, 21, 23, 31, 40]
SST k-ω	RANS	2	[33, 40, 78, 79]
Spalart–Allmaras	RANS	1	[32, 80]
Compressibility modified k-ε	RANS	2	[30, 81]
Thermally modified k-ε	RANS	2	[36, 37]
Multi-phase modified k-ε	RANS	2	[82]
LES	LES	/	[41, 83]
Laminar	/	/	[84]
DNS	DNS		[34]

k-ε model was also applied often in previous works, which introduces an additional term in its ε equation that significantly improves the accuracy for rapidly strained flows. The realizable k-ε model introduces a different transport equation to describe the turbulent dissipation rate and also a new formulation for the turbulent viscosity. This model is able to accurately predict the spreading rate of both planar and round jets, providing superior performance for flows involving rotation and boundary layers under strong adverse pressure gradients. The SST k-ω turbulence model was robust in modeling the compressible flow in divergent-convergent nozzles [38, 39]. It solves the near-wall region of the boundary layer by the robust k-ω model and switches to the k-ε in the free shear flow; thus it covers the advantages of both models. In addition to the above-mentioned models, some modified k-ε models for special usage were also employed in different studies, which will not be discussed in detail. The advanced Reynolds stress model (RSM) was also used before, in which the eddy viscosity approach is discarded, and independent Reynolds stress components are solved separately. It hence accounts for non-isotropic turbulent effects in flows with streamline curvature, swirl, rotation, and rapid changes of strain rate and thus leads to more accurate prediction of complex flows compared to other models. However, one disadvantage of this model is that it requires a high degree of coupling, aggravating the model convergence and significantly increasing the computational time. The convergence can be amended by starting the simulation with standard k-ε model and then switching to RSM. Besides the RANS-based turbulence model, laminar assumption on the gas flow was also applied in a previous study, but it is not recommended since it neglects the turbulent phenomenon caused by the extremely high gas velocity. Moreover, the large eddy simulation (LES) turbulence model, which directly solves the large eddies in the computation and implicitly accounts for the small eddies by using a subgrid-scale model, also can be found in literature. As a more advanced model, LES model normally requires huge computational time and powerful computers. A more accurate method, known as direct numerical simulation (DNS), can provide a more realistic prediction of the supersonic jet flow regime inside and outside cold spray nozzle than RANS-based and LES-based solutions, but has much higher requirement on computer performance [34]. Comparisons between each turbulence model were conducted by some researchers. Barron et al. compared the standard k-ε, SST k-ω model and RSM models for modeling the supersonic flow in cold spray, and found that there was little deviation among the three turbulence models in the prediction of gas regime [40]. A very similar conclusion was also drawn by Liebersbach et al. who compared the standard k-ε, realizable k-ε and LES models [41]. In this chapter, a comparison between several widely used turbulence models is also provided and compared with experimental observation by means of a pure gas flowing through a divergent-convergent nozzle. The geometry of the nozzle and the experimental Schlieren image were taken from the work of Zapryagaev et al. [42]. The comparison of the results is shown in Fig. 2.2. It can be seen that the Spalart–Allmaras and RNG k-ε model provided a relatively accurate prediction of the flow pattern, but Spalart–Allmaras seems to overestimate the Mach disk intensity. The realizable k-ε and SST k-ω model somewhat failed to capture the Mach disks in the free jet, while the k-ε model tended to smooth the shockwaves both inside and outside the nozzle. The RSM, despite not included in this comparison, provided results in good agreement with experiment in literature and was capable to resolve all flow features [21]. From these comparative studies, it can be noted that the turbulence model indeed had significant effects on the prediction accuracy.

Solution of the particle motion

In cold spraying, injected powders occupy a rather small volume fraction of the flow field (typically less than 10%). Requiring this, the one-way Lagrangian discrete phase modeling (DPM) method was predominantly used to compute the particle velocity and temperature. With this method, the interactions between particles and the effect of the particles on the gas phase was neglected. Specifically, the gas flow is solved in the first place, then particle parameters are calculated based on the resultant gas flow. Some publications however employed a two-way coupled Lagrangian approach for the calculation of the particle trajectories, which accounts for the effect of the particulate phase on the gas phase [8, 21–24, 27, 62, 85, 86]. This method is more suitable for the high particle loading flow than one-way Lagrangian method. The particle velocity and trajectories are predicted by the following drag force balance equation,

$$m_p \frac{dv_p}{dt} = C_D \rho (v - v_p) |v - v_p| \frac{A_p}{2} + F_b \tag{2.17}$$

where C_D, m_p, v, v_p, ρ, A_p, t and F_b are the drag coefficient, particle mass, gas velocity, particle velocity, gas density, particle cross-sectional area, time and body force. Generally, the body force term was not included in the drag force balance equation, but some works added the pressure gradient force to account for the effect of

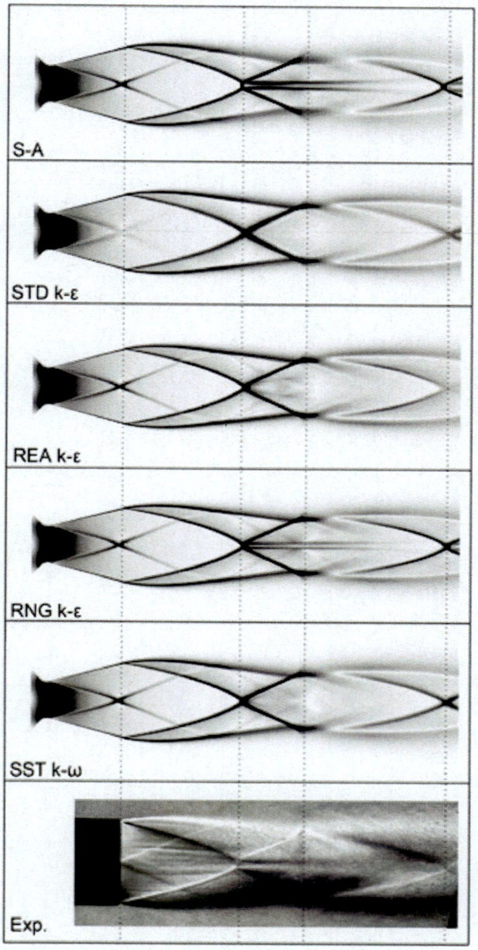

Fig. 2.2 Comparison of the gas flow pattern outside the nozzle between the CFD modeling between different turbulence models with the same working conditions. Schilieren image was taken from [42]

shock-waves (abrupt pressure changes) on the particle motion [10, 31]. Also, gravity force [72], thermophoretic force [22, 65, 77], lift force [62, 66, 77] and electrostatic force [72, 83] were included in some other studies. Besides, the Eulerian DPM was also used in a few studies [10, 82]. For this type of model, dispersed phase is considered as a continuum, and the particle velocity is calculated by solving additional discrete equations. For this reason, Eularian DPM method is more complicated than Lagrangian DPM method, but can result in more realistic results when modeling a dense particulate flow where strong coupling of particle and gas phases takes place. A detailed description and equations can be found in [82]. Moreover, in order to accurately predict the gas-particle interactions and particle spatial positions, 3D model must be applied rather than 2D [27].

Drag coefficient is the most influential factor to the particle velocity and trajectory prediction. Table 2.3 lists the commonly used drag laws for calculating the drag coefficient in cold spraying. No. 1 drag law was originally developed for laminar

Table. 2.3 Summary of the drag coefficients used for calculating the particle velocity in cold spraying

No	Authors	Equations	Range	Application
1	Schiller and Naumann [91]	$C_D = \max\left(\frac{24}{Re_p}\left(1 + 0.15Re_p^{0.687}\right), 0.44\right)$	/	Incompressible laminar
2<a>	Morsi and Alexander[92]	$C_D = a_1 + \frac{a_2}{Re_p} + \frac{a_3}{Re_p^2}$	$0.1 < Re_p < 50{,}000$	Incompressible
3	Clift et al. [87]	$C_D = \frac{24}{Re_p}\left(1 + 0.15Re_p^{0.687}\right) + \frac{0.42}{1+4.25\times10^4 Re_p^{-1.16}}$	$Re_p < 300{,}000$ $M_P > 0.4$	Incompressible/compressible
4<b>	Haider and Levenspiel [93]	$C_D = \frac{24}{Re_p}\left(1 + b_1 Re_p^{b_2}\right) + \frac{b_3 Re_p}{b_4 + Re_p}$	$Re_p < 260{,}000$	Incompressible
5<c>	Crowe [94]	$C_D = \left(C_{D(incompress)} - 2\right)\times\exp\left(-3.07\gamma^{1/2}\left(\frac{M_P}{Re_p}\right)g(Re_p)\right) + \frac{h(M_P)}{\gamma^{1/2}M_P}\times\exp\left(\frac{-Re_p}{2M_P}\right) + 2$	$0.2 < Re_p < 10{,}000$ $0.1 < M_P < 2.0$	Compressible
6<d>	Henderson [95]	$C_{D1} = 24\left(Re_p + S\left(4.33 + \left(\frac{3.65 - 1.53\frac{T_P}{T}}{1 + 0.353\frac{T_P}{T}}\right)\times\exp\left(-0.247\frac{Re_p}{S}\right)\right)\right)^{-1} + \exp\left(-\frac{0.5M_P}{Re_p^{1/2}}\right)\left(\frac{4.5 + 0.38\left(0.03Re_p + 0.48Re_p^{1/2}\right)}{1 + 0.03Re_p + 0.48Re_p^{1/2}} + 0.1M_P^2 + 0.2M_P^8\right) + 0.6S\left(1 - \exp\left(-\frac{M_P}{Re_p}\right)\right)$ $C_{D2} = C_{D1}(1.0,\ Re_p) + \frac{4}{3}(M_P - 1)\left(C_{D3}(1.75,\ Re_p) - C_{D1}(1.0,\ Re_p)\right)$ $C_{D3} = 0.9 + \frac{0.34}{M_P^2} + 1.86\left(\frac{M_P}{Re_p}\right)^{1/2}\left(2 + \frac{2}{S^2} + \frac{1.058}{S}\left(\frac{T_p}{T}\right)^{1/2} - \frac{1}{S^4}\right) \left(1 + 1.86\left(\frac{M_P}{Re_p}\right)^{1/2}\right)^{-1}$	C_{D1}: $M_P \le 1.0$ C_{D2}: $1.75 < M_P < 1.0$ C_{D3}: $M_P \ge 1.75$	Compressible

(continued)

Table. 2.3 (continued)

No	Authors	Equations	Range	Application

$Re_P = \rho d_p \left| v_p - v \right| / \mu$, $M_P = \left| v_p - v \right| / \sqrt{\gamma RT}$

<a>. a_1, a_2, and a_3 are constants that apply over several ranges of Re from 0.1 to 50,000

<c>. b_1, b_2, b_3 and b_4 are constants that apply for non-spherical particles

<c>.$C_{D(incompress)}$ is the incompressible drag coefficient proposed by Clift et al.; $\log_{10} g(Re_p) = 1.25\left(1 + \tanh\left(0.77 \log_{10} g(Re_p) - 1.92\right)\right)$; $h(M_P) = \left(2.3 + 1.7(T_p/T)^{1/2}\right) - 2.3 \tanh\left(1.77 \log_{10} M_P\right)$

<d>.$S = (\gamma/2)^{1/2} M_P$

flow, thus it should not quite proper for modeling the particle acceleration in cold spraying. No. 2 and 3 drag laws with simple expression have been integrated in the commercial software, Ansys-Fluent [87], thus they were widely used to predict the particle velocity in cold spraying. No.4 drag law is applied to non-spherical particles. Although these relations were widely used in previous modeling work, they were originally developed for the incompressible flow and hence do not consider Mach number effects. Note that No.2 incorporates a correction for particle Mach numbers exceeding 0.4 and can hence be applied to compressible regimes. Nevertheless, it is invariant under Mach number changes, which limits its applicability to the acceleration process. As reported in a previous study [88], in particle-laden flows where the particle Mach number exceeds 0.6, shock patterns may form on the particles, which significantly affects the particle motion. For a typical cold spray process, the gas velocity in the divergent part of the nozzle and in the jet is supersonic. As the particle speed is noticeably below the gas velocity, the relative fluid flow can be expected to experience such compressibility effects. Therefore, it is better to use those drag laws which include a Mach number term to consider this aspect. No. 5 and 6 drag laws are such kind of coefficients which were developed for compressible flow around spheres. Some rarely used drag law also can be found in previous publications, which we will not discuss in detail [23, 44, 84, 89, 90]. Some studies also compared different drag laws and investigated their effects on particle acceleration and heating [33, 67, 89, 90]. Ning et al. compared No. 2 and No. 6 drag laws and found that No. 6 drag law led to higher particle velocity than No. 2 [67]. Breuninger et al. compared No. 2 and No. 3 drag laws and found a good consistency when powders had a size of smaller than 5 μm but a larger discrepancy for larger particles [33]. In the studies of Leitz et al. various drag laws available in the opensource CFD code (i.e. Openform) were compared, and the results clearly demonstrated that the selection of a proper drag law was critical for the accurate modeling of particle velocity and temperature in cold spray [89, 90]. It is therefore plausible to conclude that drag law is a critical factor that can influence the modeling accuracy of cold spray particle acceleration and heating. In addition, it is also worthy to point out that for sub-micro particles, a Stokes–Cunningham correction factor was normally introduced in the drag law equations to consider the slip at the interface between the particle and the flow [66, 73, 83]. Also, particle dispersion due to turbulence effect is always considered by a stochastic tracking model.

Solution of the particle heating

Before discussing the particle heating, Boit number must be examined in the first place. It represents the ratio of internal conduction and external convective resistances to heat transfer, given as $Bi = hd/\lambda_p$ where h is the convective heat transfer coefficient, d is the particle diameter and λ_p is the heat conductivity of the spray material. Having a Boit number smaller than 0.1, the temperature within the material can be regarded as constant. The Biot number of cold sprayed particles has been proved to be much smaller than 0.1 [7, 8, 32, 96, 97]. As such, the particle temperature in cold spraying was normally assumed as homogeneous within individual particles. Therefore, the following heat transfer equation is used to calculate the particle temperature [87],

$$m_p C_{particle} \frac{dT_p}{dt} = A_S h(T_r - T_p) \quad (2.18)$$

where $C_{particle}$, T_r, T_p, A_s, h is the particle specific heat, recovery temperature, particle temperature, particle surface area and convective heat transfer coefficient, respectively. The recovery temperature is dependent on the particle Mach number and calculated by the following relation, $T_r = T\left(1 + r\frac{\gamma-1}{2}M_P^2\right)$, where r is the recovery coefficient close to 1 in gases, T is the local gas temperature and M_p is the particle Mach number. In addition, the heat transfer coefficient was normally calculated from the Nusselt number using Ranz-Marshall correlation (Eq. 3.6) which is suitable for the flow past a spherical particle [98].

$$Nu = \frac{hd_p}{\lambda} = 2 + 0.6Pr^{0.33}Re_p^{0.5} \quad (2.19)$$

where Nu, d_p, λ, Pr and Re_p is the Nusselt number, particle diameter, gas thermal conductivity, Prandtl number and particle Reynolds number, respectively. Some modified forms of the Ranz-Marshall correlation or other correlation for Nusselt number were also used occasionally in previous works for the consideration of the high particle Reynold number [99], high Mach number [100] or boundary layer on the particle surface [101]. Despite several correlations for Nusselt number have been used, a comparison between each equation is lack so far, thus it is still not clear which equation is more accurate. A comparing work may be needed in the future woks.

Solution of heat transfer between gas flow and solid walls

Heat transfer between the gas flow and solid walls including nozzle wall and substrate can be considered by CFD modeling. The region where solid phase occupies should be meshed and given properties, then the solid and gas phases can be thermally coupled through the following equation [102],

$$q = \lambda_{solid}\left(\frac{\partial T_{solid}}{\partial n}\right)_w \quad (2.20)$$

where q, λ_{solid}, and $\left(\frac{\partial T_{solid}}{\partial n}\right)_w$ are the heat flux through the interface, solid thermal conductivity and temperature gradient at the wall surface. Heat transfer through the interface can be directly calculated from the solution in the adjacent cells [102]. The boundary between the solid wall and atmosphere is normally set as heat-insulated wall. The heat transfer process within the solid walls is governed by the energy equation for steady-state heat conduction which is given as follows:

$$\frac{\partial^2 T_{solid}}{\partial x} + \frac{\partial^2 T_{solid}}{\partial y} + \frac{\partial^2 T_{solid}}{\partial z} = 0 \quad (2.21)$$

2.2.2 Gas Dynamics Properties in Cold Spraying

General gas dynamics

The acceleration of the gas velocity inside the cold spray nozzle mainly happens at the throat section where the flow transits from subsonic to supersonic. When gas comes to the divergent part, its velocity keeps increasing continuously until the near-exit region where complex wave structures may generate due to the pressure adaption. Figure 2.3 shows three typical flow structures (represented by velocity contours) of the gas inside the nozzle divergent part and outside the nozzle without the appearance of substrate, namely, under-expanded, correctly-expanded and over-expanded [101]. For the under-expanded flow shown in Fig. 2.3a, the first appearing structure is expansion waves as a result of higher nozzle exit pressure than ambient pressure, following by the shockwaves as a result of the multiple reflections at the jet boundary, and then two kinds of waves appear periodically in the jet. As a consequence, the gas velocity fluctuates periodically in the form of a sharp increase followed by a rapid decrease. Correspondingly, over-expanded flow shown in Fig. 2.3c gives a reverse

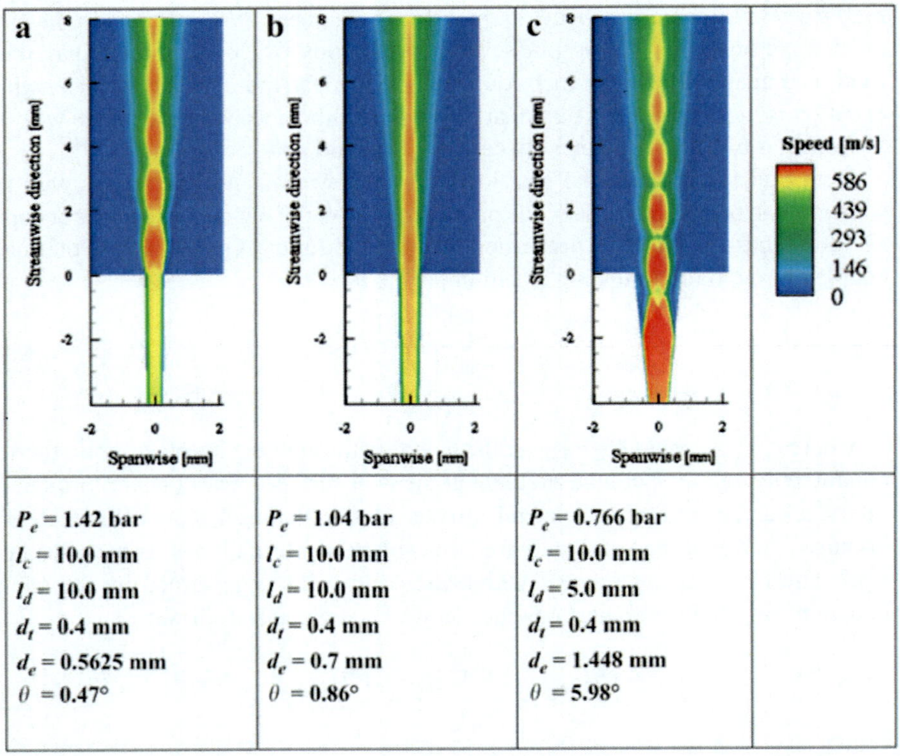

Fig. 2.3 CFD modeling result of the gas velocity contour inside the nozzle divergent section and outside the nozzle without the substrate at the ambient pressure of 0.1 MPa: **a** under-expanded, **b** correctly-expanded and **c** over-expanded [101]

structure. If the gas is correctly expanded that the pressure at the nozzle exit has no significant difference from the ambient pressure, wave structures may be greatly reduced as shown in Fig. 2.3b. No matter which flow structure forms at the nozzle exit, the bow shock enclosing a high-density and low-velocity region will be generated on the substrate surface when the supersonic gas jet finally impacts on the substrate. The gas velocity suddenly reduces from supersonic to zero inside the bow shock.

Effect of the working conditions on the gas velocity

Numerical investigation on the effect of main gas conditions (stagnation pressure, stagnation temperature and gas species) on the gas flow properties have been conducted in many studies. For the stagnation pressure, 1D isentropic theory (Eqs. 2.1 and 2.5) tells us the gas velocity in the nozzle is independent to the stagnation pressure as shown in Fig. 2.1a. However, in the same figure, CFD modeling showed that gas velocity at the nozzle exit increased with increasing the stagnation pressure and gradually leveled off [9]. The reason for the difference is because the 1D isentropic theory does not consider the effect of ambient pressure. In practice, stagnation pressure indeed influences the flow inside the nozzle because of the ambient pressure [9, 14, 69]. Figure 2.4 illustrates a schematic picture of the flow through a de-Laval

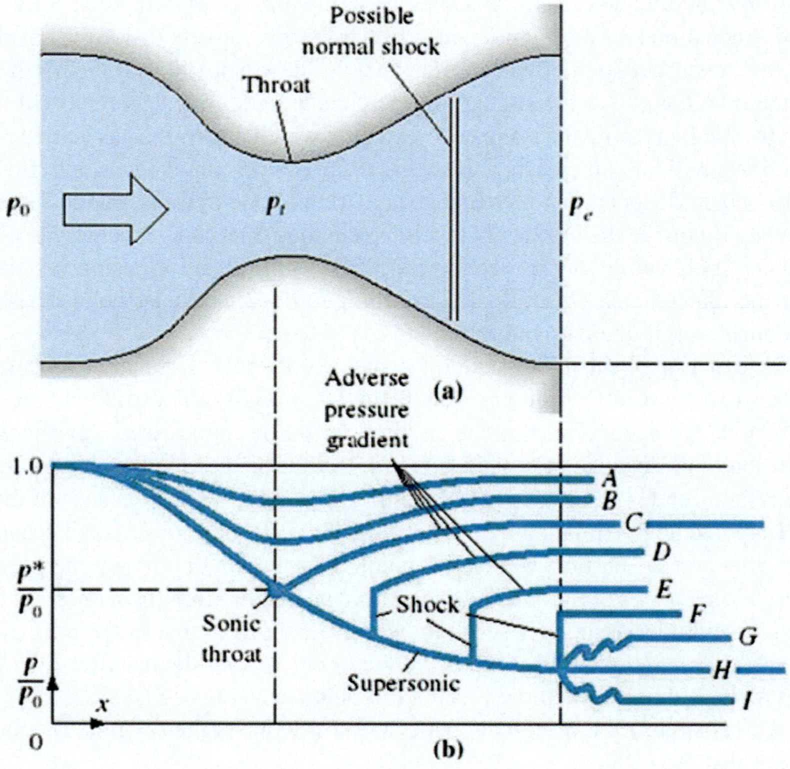

Fig. 2.4 Schematic picture of the flow through a de-Laval nozzle [84]

nozzle to explain how stagnation pressure affects the gas flow [84]. Note that the ambient pressure in normal cold spray system is constant and equal to the chamber pressure (atmosphere pressure). Therefore, the following discussion is based on the assumption of a fixed ambient pressure. Firstly, if the stagnation pressure is so low that the critical condition for achieving sonic at the nozzle throat is not satisfied, the gas velocity will stay subsonic in the entire nozzle (A and B). An extreme case is that the critical condition is achieved and thus gas is able to reach sonic at the throat, but then it suddenly reduces to subsonic again at the divergent part (C). Note that A, B and C situations do not take place in cold spray process. Secondly, as the stagnation pressure increases gradually, gas starts to be supersonic in the divergent part. However, normal shockwaves will generate in the divergent part, which results in the sudden reduction of the gas velocity from supersonic to subsonic across a shock system (D, E and F). Thirdly, with further increasing the stagnation pressure, normal shockwaves disappear. Instead, gas keeps supersonic in the divergent part until the nozzle exit region where the aforementioned oblique shocks and expansion waves generate (G, H and I). Finally, as further increases the stagnation pressure, the outside jet remains under-expansion (I). In this case, the ambient pressure cannot affect the flow inside the nozzle anymore, which means the relation between the stagnation pressure and the flow properties inside the nozzle will roughly follow the 1D isentropic theory and thus the gas velocity at the nozzle exit will keep constant. In a word, stagnation pressure significantly affects the gas velocity inside the nozzle when at a low level, but such effect becomes negligible when it is sufficiently high. But it should be noted that the strength of the outside under-expansion jet will increase with further increasing the stagnation pressure, which means the gas velocity outside the nozzle will keep increasing. Besides, from 1D isentropic equations (Eq. 2.5), it is also easily deduced that increasing stagnation temperature is able to increase the gas velocity inside the nozzle. With CFD technique, the same conclusions were also drawn, which can be clearly seen in Fig. 2.1b [9]. In addition, using helium as the main gas also led to higher velocity according to Eq. 2.5 because helium has smaller molecular weight than air and nitrogen [25, 66, 67].

The situation of substrate (standoff distance, substrate shape and substrate angle) was also found to affect the gas flow feature, especially the nozzle outside jet and bow shock. It was reported that the gas pressure in the compressed bow shock varied as the standoff distance increased [31, 82]. But the size of the bow shock showed a decreasing trend both in width and height as shown in Fig. 2.5 because of the shortened supersonic potential core of the outside jet [29]. As the bow shock has deceleration effect on the particle velocity, a detailed discussion on this will be provided in the following text. For the substrate shape and angle, the strength or shape of the bow shock on cylinder or angular substrate was different from that of the normal-placed flat substrate. Cylinder substrate or conical cold spayed deposits resulted in less-strength bow shock as the flow was not fully blocked by the substrate [31, 77], while angular substrate led to irregular-shape bow shock due to the asymmetric placement of the substrate [45].

Powder injection conditions as another influential factor to the gas flow feature were also investigated. It was found that the gas velocity and temperature inside

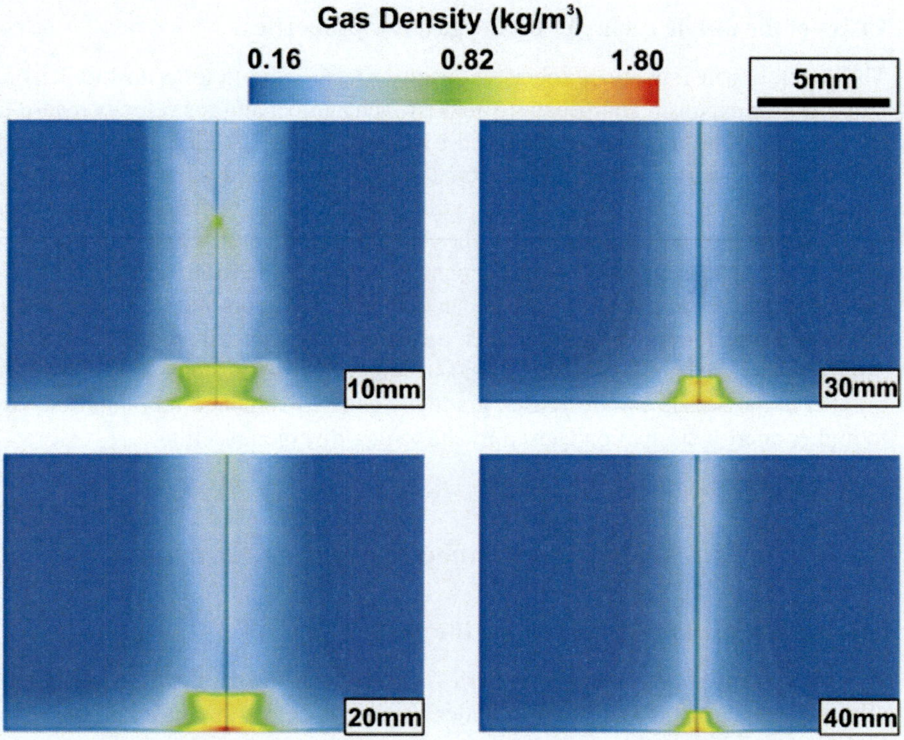

Fig. 2.5 CFD modeling result of the impinging jet and the bow shock in front of the substrate surface represented by gas density at different standoff distances [29]

the nozzle decreased as the injection pressure increased due to the heat exchanges between the low-temperature carrier gas and high-temperature main gas [65, 68]. Conversely, higher injection temperature was found to play a positive role for the gas velocity and temperature because of the extra energy input from the carrier gas stream to the main gas stream [70]. Moreover, helium as the injection gas led to a higher gas velocity inside and outside the nozzle than air and nitrogen [71]. Besides, it was also reported that putting the injection position in the pre-chamber resulted in more uniform velocity and temperature distribution in the nozzle [46, 65].

In many of previous modeling works, particle loading effects on the gas phase were neglected because the volume fraction of particles was below 10–12%. However, it was reported in some studies that increasing the powder feed rates was able to lower the gas momentum and thus weaken the shocks, expansion waves and bow shock [21, 27, 62, 82, 85, 86]. For the same reason, the gas velocity also decreased as the powder rates increased. These facts suggest that the interaction between gas and particular phases must be taken into account, especially when high powder feed rate is employed.

Effect of the nozzle geometry on the gas flow properties

The nozzle expansion ratio is a decisive design aspect for a cold spray nozzle. Theoretically, higher expansion ratio leads to higher Mach number and gas velocity regardless of the stagnation pressure as described by Eqs. 2.1 and 2.5. However, in practice, stagnation pressure significantly affects the flow inside and outside the nozzle as discussed before. Only when the stagnation pressure was sufficiently high so that ambient pressure no longer affected the flow inside the nozzle, the Mach number and gas velocity at the nozzle exit augmented with the expansion ratio [14, 69]. This fact indicated that large expansion ratio nozzle was able to produce high gas velocity but required suffciently high stagnation pressure at the same time [14]. Besides, the nozzle divergent length was also found to affect the gas velocity [14]. The Mach number at the nozzle exit decreased gradually as the divergent length increased due to higher energy dissipation caused by the longer divergent section.

2.2.3 *Particle Velocity and Trajectory in Cold Spraying*

Effect of the working conditions on the particle velocity

Particle acceleration behavior is primarily dependent on the gas flow features. Generally, higher gas velocity results in better particle acceleration and the consequent higher particle velocity. So far, it has been accepted that the particle velocity increased with the increment in gas stagnation temperature, stagnation pressure or using helium as the main gas because of the increased gas velocity or density inside and outside the nozzle [3, 7, 25, 45, 49, 50, 61, 66, 67, 74, 103], which can be clearly seen in Fig. 2.6 [74]. It must be known that increasing stagnation temperature is more influential than increasing pressure, but higher gas temperature may increase the risk of

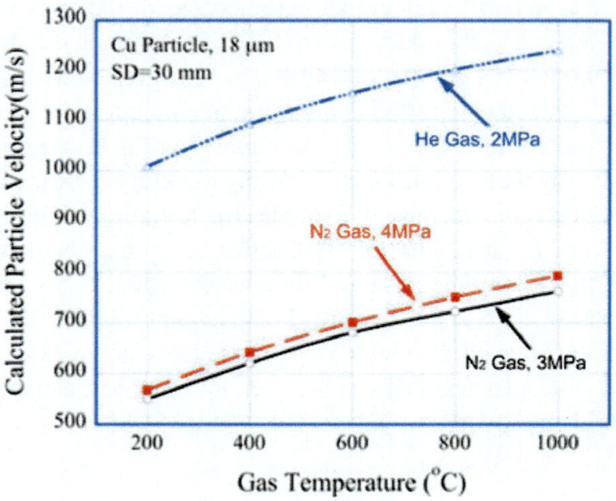

Fig. 2.6 Modeling result of the particle velocity at the location 30 mm away from the nozzle exit as a function of gas stagnation pressure, temperature and species [74]

thermal-induced defects in the coating. In addition, using helium for spraying significantly increased the economic cost. The above-mentioned conclusions were already validated by experimental measurements (See Table 2.3) and have been widely used in practice to improve the particle velocity and coating performance. Moreover, injection gas pressure, temperature, and species were also found to influence the particle impact velocity in recently published papers. Lower injection pressure, higher injection temperature or using helium as the injection gas contributed to the improvement of the mixing gas (carrier gas and main gas) velocity and thus the particle velocity upon impact [68, 70, 71].

Standoff distance as another important parameters were found to affect the particle impact velocity and thus coating deposition efficiency. Only few publications on this topic with CFD modeling can be found in literature. As discussed before, in some studies, it was proposed that the pressure in the bow shock region varied with increasing the standoff distance; the substrate should be placed at the location where the lowest pressure was achieved because the pressure inside the bow shock determined the particle deceleration. However, Pattison et al. proposed that it was the combination of outside supersonic jet core length and bow shock determined the particle velocity. Based on this, they reported that there existed an optimal standoff distance to guarantee the maximum particle velocity and coating deposition efficiency [29], which was validated by experiment. Another different result was given by Li et al., which showed that coating deposition efficiency and thus particle velocity upon impact roughly decreased as the standoff distance increased [56]. The different results reported in these publications tell us that the role of standoff distance in particle velocity upon impact and deposition efficiency is still not quite clear. More works on the standoff distance should be carried out in the future. On the other hand, substrate geometry and placed angle also influenced particle velocity, particularly the normal velocity component which determines the deposition efficiency [31, 45]. In contrast to the flat substrate placed vertically to the nozzle axis, the particles impinging on the cylinder or angular substrates attained lower normal velocity component due to the radius of curvature and angle, which had a negative effect on the coating deposition.

Particle properties represents another factor that affects particle velocity. It was seen from several publications that the particle velocity at the nozzle exit decreased as particle size (or density) increased, because lighter particles were more susceptible to the supersonic driving gas as describe by Eqs. 2.6 and 2.7. For the same reason, when these particles penetrated into the bow shock region, their velocity abruptly reduced to a very small value despite they got high velocity at the nozzle exit. On the contrary, for heavier particles, the velocity at the nozzle exit was small, but they decelerated only slightly across the bow shock [7, 19, 33, 66, 68, 70, 76, 77, 84, 103–106]. Figure 2.7 shows particle velocity upon impact as function of copper particle size calculated by CFD modeling and analytical equations [19]. As discussed above, smaller particles lost much velocity after the bow shock, while considerably larger particles exhibited a low velocity before reaching the bow shock. As a consequence, the maximum velocity upon impact was achieved at an intermediate size. Further increasing or decreasing the particle size resulted in a respective reduction of their velocity. This relation is not only valid for copper but for all metals. In addition, it

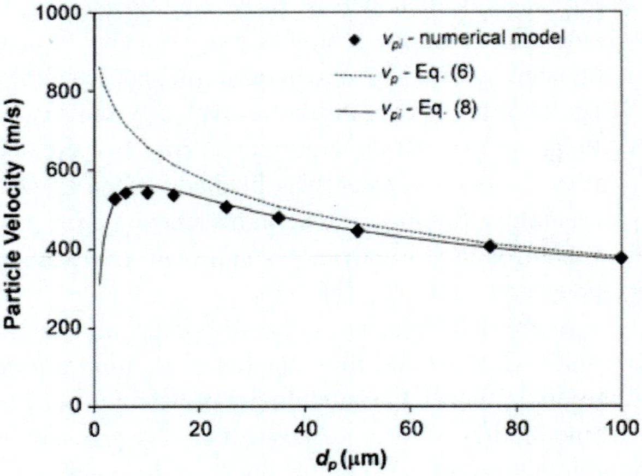

Fig. 2.7 Particle velocity upon impact as function of copper particle size predicted by CFD modeling and analytical equations [19]

was also reported that irregular particles experienced a stronger acceleration than spherical particles due to the larger drag force. Hence, under the same equivalent size, irregular particles had higher impact velocities [57, 60].

Particle feed rate was found to have an effect on the gas flow and thus particle acceleration under high particle loadings. It was reported that particle velocity had a slight decrease as the particle loading increased at low feed rates [24, 25], which means the momentum exchange between gas and particle phase can be ignored. However, when powder loading rate was high, such effect became significant and the velocity reduction became prominent with increasing the feed rate [16, 27, 82].

Effect of the nozzle geometry on the particle velocity

In terms of nozzle geometric, the expansion ratio strongly affects the particle velocity. Nozzles with higher expansion ratio result in higher particle velocity but required higher stagnation pressure which corresponds to the gas properties. For a specific nozzle with fixed stagnation pressure, there was an optimal expansion ratio which guaranteed the maximum particle velocity [14, 48, 49, 63]. Further decreasing the expansion ratio led to low gas velocity, while increasing the expansion ratio led to the flow inside the nozzle affected by the ambient pressure. Both facts will result in the decrease of the particle velocity. Besides, nozzle divergent length was found to pose significant impact on the particle impact velocity [14, 49]. Figure 2.8 shows the particle velocity upon impact as a function of nozzle divergent length [14]. As displayed, with increasing the divergent length, the particle velocity grew at first due to the increased acceleration time in the divergent part. However, at a certain length, particle velocity turned to decrease with further increasing the divergent length because gas velocity became so low that particle started to decelerate. Studies also reported that the ratio of the throat width to the divergent length for a rectangular nozzle posed significant impact on the viscous boundary layer in the nozzle divergent section and thus particle acceleration under a constant expansion ratio [107]. A ratio of less than 0.02 was not recommended for rectangular nozzle design due to the

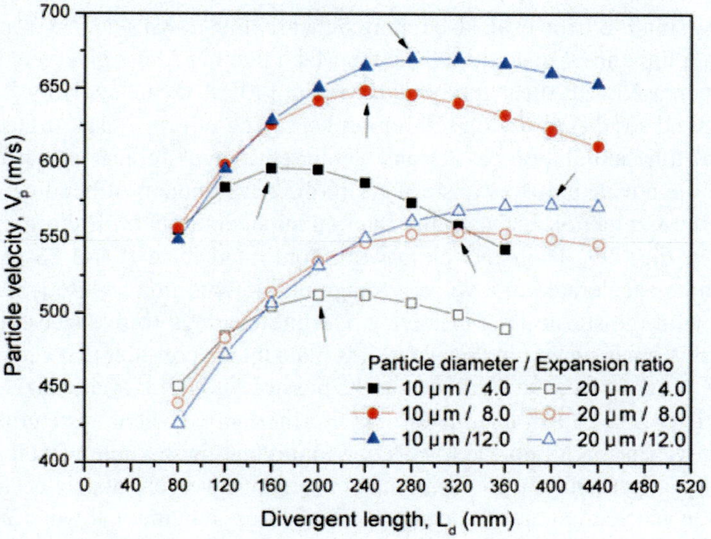

Fig. 2.8 Modeling result of particle velocity upon impact as a function of nozzle divergent length with different particle sizes and expansion ratios [14]

strong boundary layer effects which led to a reduction of particle velocity in the nozzle [17, 107]. To date, a similar investigation on the round-shape nozzle has not been carried out. To sum up, an optimal design of the nozzle should firstly determine the expansion ratio according to the designed stagnation pressure and then identify the adequate divergent part length and cross-section shape to achieve the maximum particle impact velocity.

The powder injection location also plays important role in the particle acceleration behavior. For particles releasing in the high-pressure section, the injection position rarely affected the particle velocity because the particle acceleration mainly took place in the divergent part [51]. However, if the injection point was in the low-pressure part of the nozzle, the impact velocity decreased as the injection position moved toward the nozzle exit due to the shorter acceleration time [51]. Besides, in some cases, particles were injected from the pre-chamber to the nozzle. By doing this, particle temperature were improved because of the longer heating time in the chamber [46, 65]. Interestingly, particle velocity were also found to increase as the preheating chamber length increased [65]. However, one problem is moving the injection point far from the nozzle inlet may lead to serious particle dispersion. Therefore, whether this is a good way still needs further experimental validation.

Some special-designed nozzles were also tested with CFD technique to estimate its potential performance. Micro-scale nozzle with the dimensions far less than the conventional cold spray nozzle is a useful tool to fabricate the coatings on the component inner surface or employed in the vacuum cold spraying. Existing works showed that such nozzle also produced supersonic jet and high particle velocity [40, 78, 80], but particles must had micro-size or nano-size because the nozzle throat dimension was very small. Barron et al. also reported that scaling down a nozzle from its normal size to a miniature size could reduce the Mach number in the divergent section due to the increased wall shear and boundary layer [40]. A recent study by Kiselev et al. investigated a radical nozzle used for cold spraying of pipe inner surface too [108]. The real-time simulation results and Schlieren images indicated that the jet flow outside the nozzle exit was unsteady with the development of bending oscillations due to the interaction between the internal supersonic jet with the nozzle wall as shown in Fig. 2.9. Their modeling work confirmed that 10 and 25 μm Al particles can be accelerated to a velocity range of 345–405 m/s and 360–390 m/s upon impact with the substrate respectively. Despite coatings were successfully formed on the pipe inner surface in this work, it is plausible to consider that a higher particle velocity may not be achievable due to the possible clogging of the nozzle caused by the particle–wall collision at the turning. Furthermore, Li et al. tested the feasibility of a convergent-barrel nozzle [47]. They found that the particle velocity was much lower in comparison to a conventional convergent-divergent nozzle because the gas cannot be accelerated at the barrel part. In addition, a numerical work also reported that a bell-shape nozzle could result in higher particle velocity than a conical-shape nozzle with a same expansion ratio. However, the conclusion cannot be fully validated due to the different lengths of the two nozzles used for comparison and lack of experimental validation [61]. Other new techniques such as liquid feedstock [23] and shock-tube-assisted cold spray [109] were also tested by CFD modeling. Here we will not discuss in detail.

Particle trajectory and spray spot footprint

As a preliminary study, Karimi et al. [30] developed a CFD model which consisted of two sub-models (nozzle domain and environment domain) to explore the particle trajectories and distribution in the gas stream. In that study, particles were injected to the nozzle at an angle of 45° rather than in the axial direction, which were hence found to concentrate at one side of the substrate. Later on, a complete 3D model was

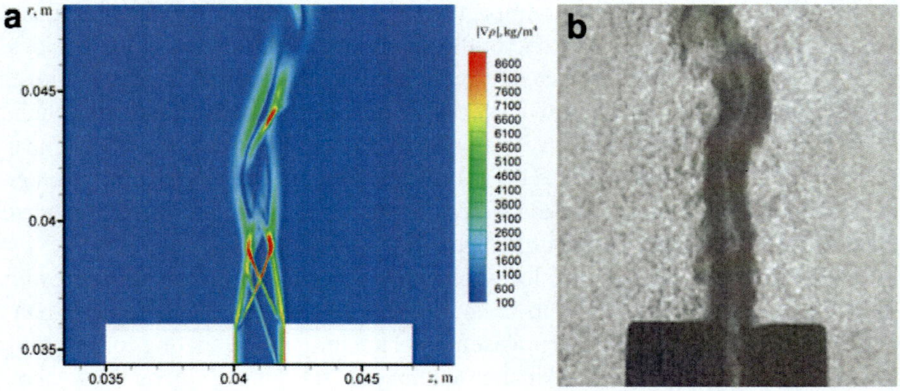

Fig. 2.9 Comparison of the real-time simulation results and Schlieren photograph of a jet flow outside the exit of the radial nozzle [108]

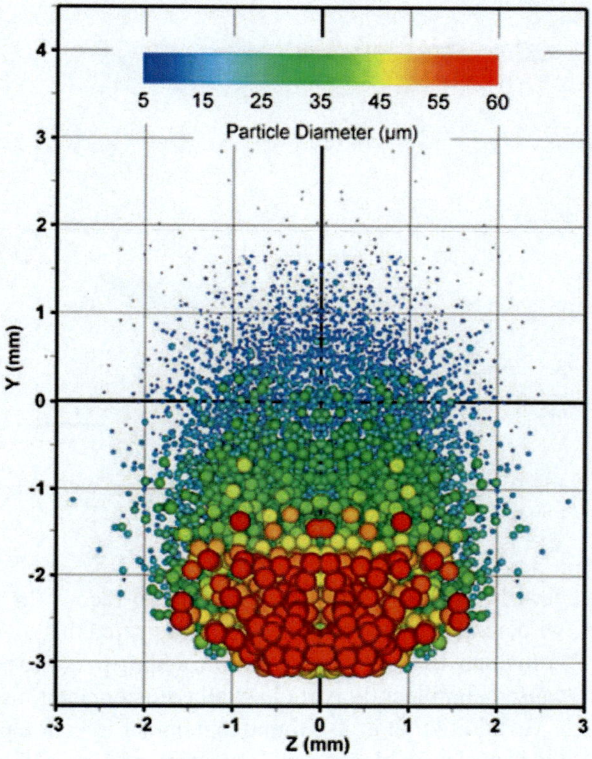

Fig. 2.10 Footprints for the particles ranging between 5 and 60 μm on a flat substrate 10 mm away from the nozzle exit [31]

developed based on a similar nozzle configuration in order to predict the effect of particle size on the footprints [31]. Figure 2.10 shows the footprints for the particles ranging between 5 and 60 μm on a flat substrate [31]. The results showed that large particles mainly concentrated at the center-bottom region, whereas small particles dispersed more seriously because they were more susceptible to the gas flow.

The above-mentioned two studies mainly focused on the radial injection nozzle. For the widely used axial injection nozzle, it was reported that particles dispersed at the divergent section and out of the nozzle exit with a higher particle concentration in the central region [37, 41, 46, 68, 75, 76, 110]. Higher gas pressure and temperature reduced the extent of dispersion and narrowed the of footprints [58]. Figure 2.11 shows the trajectories of titanium particle stream colored by velocity magnitude inside and outside the nozzle [37], where particle dispersion and collision with the nozzle wall can be clearly observed. Such dispersion resulted in the collision between particles and nozzle wall, increasing the potential risk of nozzle clogging. A study of Ozdemir et al. further demonstrated that the dimensions of injector could influence the particle–wall collision in the nozzle and thus nozzle clogging [110]. The powder feeder with larger diameter was more likely to cause particle–wall collision at the nozzle divergent section, and a slight misalignment of the injection direction from the central axis would lead to severe particle–wall collision. Furthermore, in Liebersbach

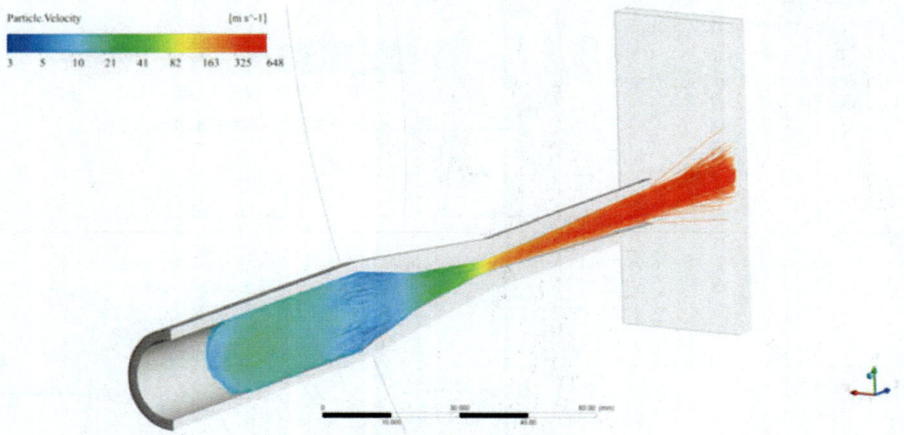

Fig. 2.11 Trajectories of titanium particle stream colored by velocity magnitude at the stagnation pressures and temperature of 1.4 MPa and 550 °C [37]

et al.'s study, a UDF was developed to record the particle velocity upon collision with nozzle inner wall. The results indicated that under steady state the particle–wall collision was hardly happen, while adding pressure oscillations for the powder feeder caused significantly particle–wall collision and thus possible nozzle clogging [41].

In addition, it is also found that higher injection pressure helped the dispersion of the particles inside the nozzle convergent part [37]. In addition, spatially, the particle velocity was found to decrease gradually in the radial direction outwards from the center [10, 25, 58, 68, 75, 76] due to the corresponding gas velocity distribution at the cross-sectional planes. For the same reason, it was suggested to use rectangular and circular nozzles for cold spraying because more high-velocity particles were at the central region [25, 51, 52]. A recent study even built a profile of cold sprayed deposits based on the CFD prediction of particle footprints, velocities and temperatures [76].

2.2.4 Particle Temperature

Particle temperature was found to influence the critical velocity; higher particle temperance upon impact was favorable for reducing critical velocity and the consequent coating formation [19, 104, 111]. Therefore, it is worthy to investigate the particle temperature. Note that experimental measurement on the particle temperature is still lack so far due to the technique limitation; existing studies were mainly on the basis of CFD modeling and theoretical analysis. It was reported that higher stagnation temperature facilitated the particle heating due to the higher heat energy input from the gas phase [7, 50]. Besides this, to increase the particle temperature must scarify particle velocity because of the energy conservation of the gas flow (high gas velocity means low gas temperature). This fact means that one cannot increase both particle velocity and temperature at the same time. For the purpose to achieve both high particle velocity and temperature at the same time, preheating chamber before the nozzle was developed and widely applied. The modeling results indicated the releasing particles in the preheating chamber was able to result in high particle temperature [37, 46, 72]. Taken overall, higher stagnation temperature or using preheating chamber are two effective ways to simultaneously obtain high particle velocity and temperature.

The relationship between particle temperature, size and critical velocity was quantified by Assadi et al. through an empirical equation, which is given as follows [19].

$$v_{cr} = v_{cr}^{ref} \frac{0.42\left(d_p/d_p^{ref}\right)^{0.5} \sqrt{1 - T_p/T_m} + 1.19\sqrt{1 - 0.73T_p/T_m}}{0.65 + \left(d_p/d_p^{ref}\right)^{0.5}} \quad (2.22)$$

where v_{cr}, v_{cr}^{ref}, d_p, d_p^{ref}, T_p and T_m the critical velocity, reference critical velocity (650 m/s), particle diameter, reference particle diameter (10 μm), particle temperature upon impact and particle melting point. Based on Eq. 3.9, one can plot a curve of temperature-dependent critical velocity against particle size. If the corresponding particle velocity upon impact against particle size is also plotted in the same figure, a deposition window for proper particle size range can be obtained. Figure 2.12 shows the schematic of the deposition window [105]. This deposition window helps the selection of working parameters and particle size in the real cold spray experiment and application.

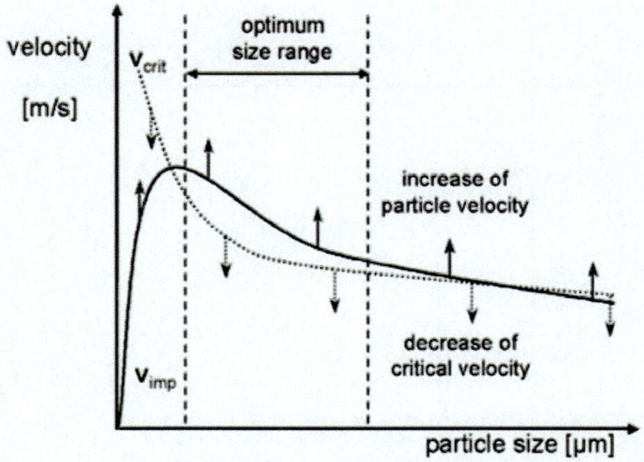

Fig. 2.12 Schematic of the deposition window (optimum particle size range) in terms of particle size. Solid line donates the particle velocity upon impact and dash line donates the corresponding critical velocity [105]

2.2.5 *Heating of the Substrate and Nozzle Wall*

Substrate and nozzle wall heating behavior were investigated as they were found to affect the final coating deposition. Higher substrate temperature helped the coating formation [112–114], while lower nozzle wall temperature reduced the nozzle clogging risk and thus increased deposition efficiency [64]. The first study regarding to the substrate temperature was reported by Kosarev et al., which presented a preliminary result of the temperature distribution within the substrate by solving the 2D heat-transfer equations [18]. Whereas a CFD study on this topic was firstly carried out by Yin et al. [59]. It was found that the substrate temperature decreased gradually from the central region towards the edge and back surface, which was consistent with both theoretical results and experimental measurements [33, 59]. Besides, it was reported that higher stagnation temperature or using materials with smaller conductivity contributed to the substrate heating [53]; standoff distance, substrate size and spray angle also had effects on the temperature distribution but not seriously [54, 115, 116]. Moreover, the investigation was extended to the numerical model design for the improvement of the prediction precision. Figure 2.13 shows the comparison of the substrate temperature between the modeling results with original k-ε turbulent model, with modified k-ε turbulent model and experimental measurement. As

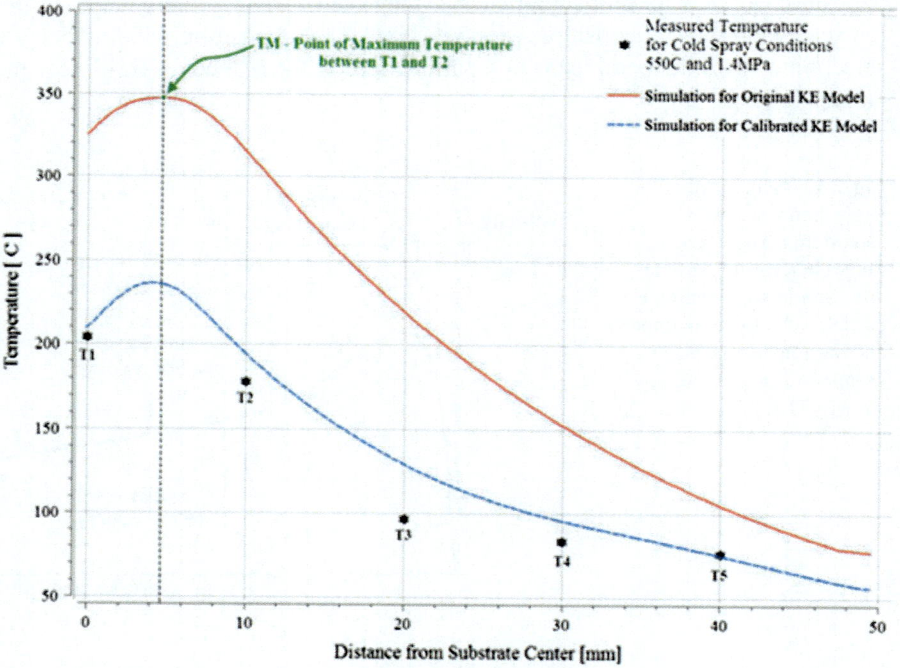

Fig. 2.13 Comparison of the substrate surface temperature between the CFD modeling with different k-ε turbulence models and experimental measurement at the inlet pressure and temperature of 1.4 MPa and 550 °C [36]

illustrated, the modified k-ε turbulence model resulted in more accurate prediction than original k-ε turbulence model which greatly over-predicted substrate surface temperature [36].

The temperature distribution of the nozzle wall was also investigated with CFD method. It was found that the high-temperature zone was mainly located at the near-throat region [53, 64]. Such high temperatures caused nozzle clogging when spraying low-melting temperature materials, such as aluminum. Particularly, if the nozzle was made of low-conductivity materials, e.g. stainless steel, the wall temperature was very high and nozzle clogging was much easier to happen [53]. To solve this problem, an additional cooling device was developed to reduce the nozzle temperature. Figure 2.14 shows the temperature distribution along the inner nozzle wall with and without nozzle cooling device [64]. It was found that the nozzle wall temperature significantly decreased by using the cooling device. Similarly, clogging was also found to occur in the injector tube due to the high-temperature main gas heating the outer injector wall [22]. In this case, a new injector configuration was developed, introducing an air gap between the main gas and the injector wall. A coating build-up experiment indicated that the injector tube clogging was prevented owing to the temperature reduction of the injector wall [22]. Furthermore, a recent study revealed that the energy loss through the convective heat transfer through the nozzle wall can be neglected during the design of a CFD model due to its small fraction in the total energy loss [40]. In other words, the nozzle wall can be reasonably considered as adiabatic wall. In some works, the mass growth of cold sprayed deposits were also considered to study the heat transfer between the impinging jet, cold sprayed deposits and substrate [79, 117]. The results indicated that the thermal energy rise per unit mass decreased as the powder feed rate increased due to the additional mass of the deposits absorbing the heat flow induced by the impinging supersonic jet, and thus the substrate temperature showed a slightly reducing trend with increasing the powder feed rate [79].

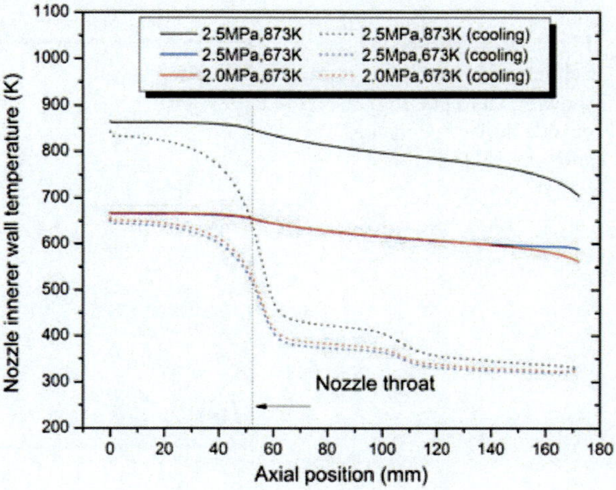

Fig. 2.14 Modelling result of temperature distribution along the nozzle inner wall with and without nozzle cooling at different working conditions [64]

2.3 Experimental Investigations

2.3.1 Visualization of the Gas Flow

Experimental observation of the gas flow is essential for the thorough understanding of the gas and particle dynamics as well as for the validation of modeling results. Visualization of the gas phase is a useful technique to analyze the flow features, which are usually difficult to detect due to the colorless gases. Currently, the most popular approach to realize the flow visualization is the Schlieren photography technique which captures the variation of the refractive index resulting from gas density gradients and transfers it into different light intensities that can be observed by human eyes and cameras. By the aid of this technique, a few works have realized the visualization of the supersonic jet flow downstream of the nozzle exit in cold spraying [18, 21, 29, 55, 108, 118, 119]. An example of the Schlieren image was already shown in Fig. 2.15 where the complex wave-structure (shock waves and expansion waves) and the highly compressed bow shock can be clearly observed [21]. For the internal nozzle flow, it is a particular difficulty to measure flow features with optical techniques, because the nozzle is normally made of non-transparent materials. Katanoda et al. applied a transparent glass as the nozzle wall to visualize the flow pattern inside a low pressure cold spray nozzle [120]. The results are shown in Fig. 2.16 where the complex shock-wave patterns are clearly observed inside the barrel. This work provided a reference for visualizing the flow inside the conventional high pressure cold spray nozzle, which has not been reported yet. Despite Schlieren photography successfully realized the visualization of the flow structures (density gradients) in cold spraying, some important parameters, e.g. velocity and temperature, cannot be quantitatively recorded by this technique. Therefore, in most cases, Schlieren photography was only used to validate the CFD models pertaining to limited phenomena of the gas phase.

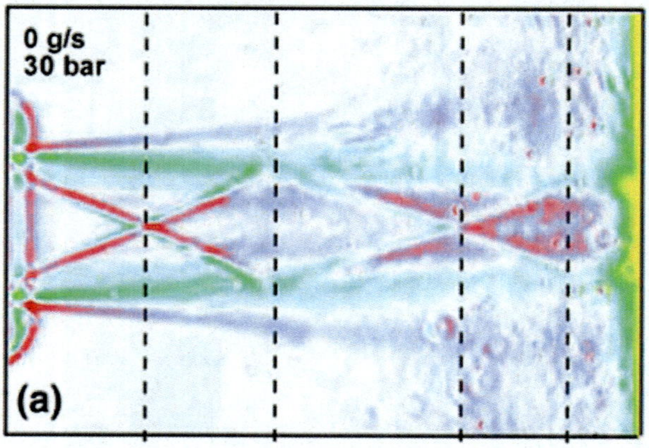

Fig. 2.15 Experimental Schlieren photograph showing an impinging jet outside the nozzle in cold spraying [21]

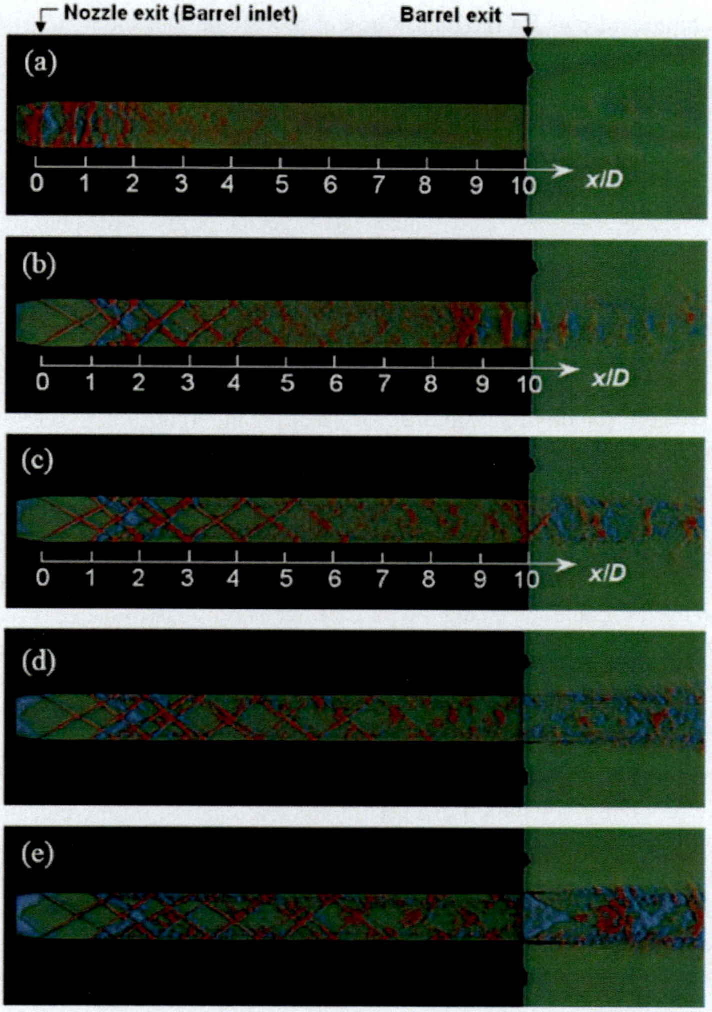

Fig. 2.16 Schilieren image of the supersonic gas flow pattern inside and outside a low-pressure cold spray nozzle with the inlet pressure of **a** 0.2, **b** 0.3, **c** 0.4, **d** 0.6 and **e** 0.78 MPa [120]

2.3.2 Particle Velocity Measurement

Experimental measurement on the particle velocity was also widely carried out over the past decades to clarify the particle motion behavior and critical velocity in cold spraying. So far, three non-intrusive techniques have been employed to measure the particle inflight velocity outside the nozzle, namely, Laser-2-focus (L2F), Doppler picture velocimetry (DPV) and Particle image velocimetry (PIV). The detailed description of these techniques can be found elsewhere [3, 10, 121]. In the early stage, L2F was used to measure the particle velocity in cold spraying [106, 121].

This technique allows for the measurement of a single particle in a limited volume, but with very low spatial resolution. Thus, it was not widely applied in latter works. DPV as a more advanced technique was frequently used nowadays due to its capability to measure the particle velocity at higher spatial resolution. Like the L2F, DPV is also suitable for single particle velocity measurement only. Nevertheless, it is still possible to get a velocity distribution map at a certain cross-section by DPV. One needs to separately measure the data at each selected point on the plane and averages the data to generate the finally recorded data. In this way, the measuring accuracy may not be guaranteed and the experiment time may be relatively long. PIV technique as another option provides a more convenient solution. It allows recording the instantaneous velocity distributions with high spatial resolution, significantly increasing the measuring accuracy and reducing the experiment time. Figure 2.17 shows the typical particle velocity maps at the cross-section and longitudinal section measured by DPV and PIV, respectively [10, 122]. A modified PIV method called particle tracking velocimetry (PTV) was developed and used for the velocity measurement of cold spray particles in recent years. In contrast with PIV, PTV uses multi-pulse laser illumination during a single camera exposure, thus PTV can provide a tracing of a specific particle flight in an image as shown in Fig. 2.18 [123]. Infrared camera was also used for characterizing the particle acceleration, which can provide a direct visualization of particle dispersion behavior outside the nozzle. But it is not capable to provide an information with regard to particle velocity [124]. A similar experimental setup which is very similar to PIV was also used to characterize the particle dispersion behavior outside the nozzle [34, 125].

Since most of the conclusions drawn from the experiment are same as those from the CFD modeling, a detailed discussion and explanation on the measuring results will not be provided in this section. Instead, a comprehensive summary of previous experimental works is made and listed in Table 2.4 where the basic particle properties and main conclusions of each work can be found. Note that inlet pressure and temperature refer to the data measured by the pressure gage and thermocouple at the nozzle inlet in experiment; in the theory analysis and CFD modeling they are

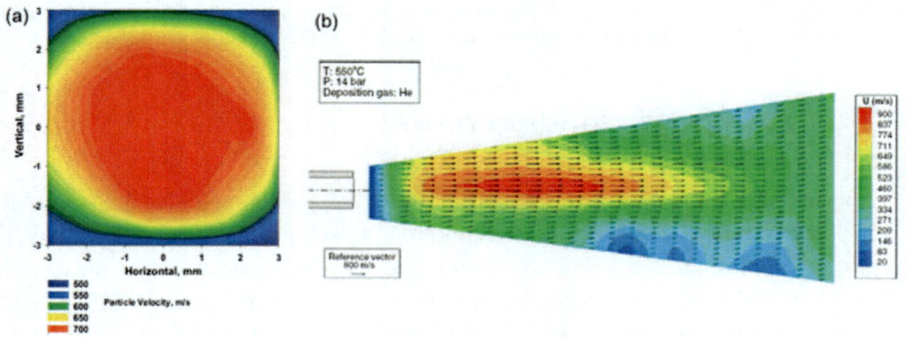

Fig. 2.17 Particle velocity distribution at the cross-section measured by DPV and longitudinal section measured by PIV in the supersonic jet outside the nozzle [10, 122]

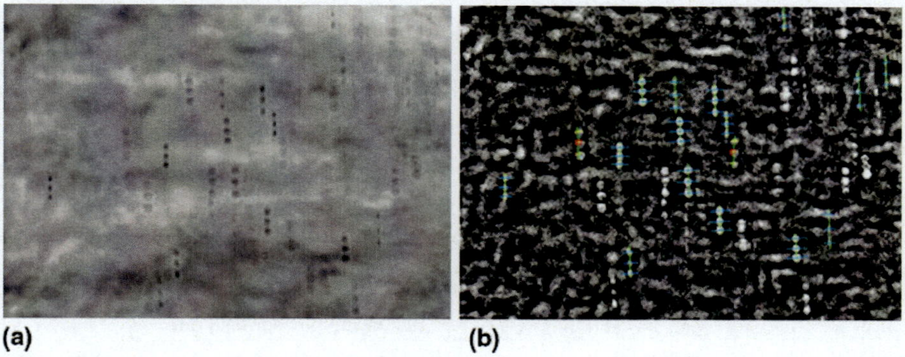

Fig. 2.18 Typical PTV image: **a** a portion of single unprocessed backlight image of the spray and **b** processed backlight image [123]

normally represented by stagnation pressure (assuming velocity at the nozzle inlet is 0) and temperature. Although most of the conclusions in Table 2.4 are in good agreement, contradictory results were also reported when studying the effect of standoff distance on the particle velocity. Pattison et al. reported that Ti particle velocity increased as standoff distance rose [29], but Zahiri et al.'s result was different [126]. Both works used Ti powders with similar morphology and size. Such contradictory results were also reported in the deposition efficiency which normally corresponds to particle velocity. Pattison et al. suggested that particle velocity increased first and then started to decease as standoff distance increased [29], which supported their own results; while the work of Li et al. showed a roughly decreasing trend [56], which supported Zahiri et al.'s results. Under current conditions, it is still not clear why different works gave such contradictory results. It seems a systematic investigation on the standoff distance is very necessary in the further works. Another interesting phenomenon were found after careful analysis of Table 2.4. The measuring results by DPV was normally higher than or equal to the CFD modeling results [10, 71, 74, 127], while PIV always led to lower value [25, 128]. So far it is not clear if this is just a coincidence or some inevitable reasons (experiment design, system error, etc.) are behind this phenomenon. Future comparison works between CFD, DPV and PIV may be needed to clarify this issue. Here it is worthy to mention two interesting works which were conducted in recent years. Koivuluoto et al. for the first time measured particle velocity and size of a cold spray particle simultaneously through a sophisticated experimental design based on the aforementioned PTV approaching, and the experimental measurement showed that particle velocity reduced as the diameter increased [123]. Another very interesting experimental study conducted by Meyer et al. measured the particle velocity inside a cold spray nozzle by using transparent quartz as the nozzle wall material [86]. Figure 2.19 shows the spatial evolution of particle velocity for Stellite-21 with increasing pressure within transparent nozzle, where particle acceleration process can be clearly seen in the nozzle.

Table. 2.4 Summary of the literature regarding to particle velocity measurement

	Materials	Shapes	Sizes, μm	Conclusions	
L2F	Cu	Spherical	19	1. Particle velocity increased with increasing gas pressure and temperature 2. Particle velocity decreased along radial direction 3. Helium resulted in higher particle velocity than air 4. Particle velocity decreased with increasing powder feed rate	[106]
DPV	Al	Spherical	15	1. CFD model validation	[9]
	Cu	Spherical	/	1. Particles footprints mainly concentrated at the central region	[129]
	316L 316L	Spherical Irregular	/ /	1. Helium resulted in higher particle velocity than air 2. Gas temperature was more influential than pressure 3. Irregular particles resulted in higher particle velocity	[130]
	Al Zn Sn	Spherical Spherical Spherical	36.2 13.6 10.6	1. Nozzle gun test	[131]
	ZK61	Spherical	58	1. CFD model validation	[71]
	Cu	Irregular	30	1. CFD model validation 2. Particle velocity increased with increasing gas temperature	[74]
	Fe alloy	Spherical	/	1. Critical velocity prediction	[132]
	Al-12Si	Spherical	/	1. Particle velocity increased with increasing gas temperature 2. Low-velocity particles account for larger share than high-velocity particles	[109]
	Al alloy Al alloy	Spherical Irregular	-25, 25–28 -25, 25–28	1. Irregular particle resulted in higher particle velocity 2. Particle velocity decreased with increasing particle size in the free jet without substrate	[133]
	BMC	Spherical	25–45	1. CFD model validation	[127]
	Al	Spherical	17–33	1. Particle size measurement 2. Initial conditions for other calculations	[97]

(continued)

Table. 2.4 (continued)

	Materials	Shapes	Sizes, μm	Conclusions	
	In718	Spherical	5–22 15–45 45–90	1. Particle velocity increased with increasing gas pressure and temperature 2. Helium resulted in higher particle velocity than nitrogen 3. Smaller particles attained higher velocity than larder particles	[134]
PIV	Cu Cu Cu	Irregular Irregular Spherical	21–62.4 20.9 12.4	1. Particle velocity decreased with increasing particle size in the free jet without substrate 2. Irregular particle resulted in higher particle velocity	[67]
	Cu Cu Al WC-12Co WC-12Co	Spherical Spherical Irregular Spherical Spherical	33 104 35 22 56	1. Rectangular nozzle was better than circular one 2. Gas temperature was more influential than pressure 3. Powder feed rate had negligible effect on particle velocity 4. Particle velocity decreased with increasing particle size and density in the free jet without substrate	[4]
	Cu Al Ti	Spherical Spherical Irregular	22.32 17.6 21.7	1. Particle velocity of Cu and Ti increased with increasing standoff distance 2. Particle velocity of Al maintained steady with increasing standoff distance	[29]
	Ti	Irregular	27	1. Helium resulted in higher particle velocity than air 2. Particle velocity decreased along the radial direction	[122]
	Al-12Si	Spherical	25	1. Particle velocity increased with increasing gas pressure and temperature 2. Particle velocity decreased along radial direction	[135]
	Ni	/	19	1. Particle velocity increased with increasing gas pressure and temperature	[3]
	Cu Ni	/	16.2	1. For critical velocity prediction	[13]

(continued)

Table. 2.4 (continued)

	Materials	Shapes	Sizes, μm	Conclusions	
	Cu	Spherical	5–25	1. Particle velocity decreased with increasing particle size in the free jet without substrate 2. Helium resulted in higher particle velocity than air 3. CFD model validation	[25]
	Cu, Zn, Al	/	/	1. Miconozzle configuration validation	[80]
	Cu Cu Cu	Spherical Spherical Spherical	5 10 15	1. Particle velocity increased with increasing gas pressure and temperature 2. Particle velocity decreased with increasing particle size (measuring position unknown)	[113]
	Ti	Irregular	22	1. Particle velocity decreased with increasing standoff distance	[126]
	316L	Spherical	22–36	1. High velocity particles was mainly at the central region 2. For validation of the CFD model	[128]
	Al	6061		1. CFD model validation	[110]
	Ti	Spherical	29	1. Particle velocity increased with increasing gas pressure and temperature 2. Helium resulted in higher particle velocity than nitrogen 3. For validation of CFD model	[58]
	Cu Nb	Spherical	25 23	1. For validation of CFD model	[89, 90]
	Cu	Irregular	10–50	1. Particle velocity increased with increasing gas pressure and temperature 2. Nozzle cross-section influenced particle velocity	[107]
	stellite-6 AL Ti	Shperical	10–60	1. Gas and particle velocities decreased with increasing powder feed rate 2. Particle velocity decreased along radial direction 3. Injector location influenced particle dispersion and velocity	[85, 86, 136]
PTV	Ni SS316L WC-SS305 Cu Cu + Al2O3	Spherical	10–30 25–10 5–25 5–25 5.6–22.5	1. Particle velocity decreased with increasing particle size (measured in single experiment) 2. Particle velocity decreased along radial direction	[123, 137]

(continued)

Table. 2.4 (continued)

	Materials	Shapes	Sizes, μm	Conclusions	
	Steel	Spherical	11–45	1. Particle velocity decreased along radial direction 2. Particles dispersed after leaving the nozzle	[124]

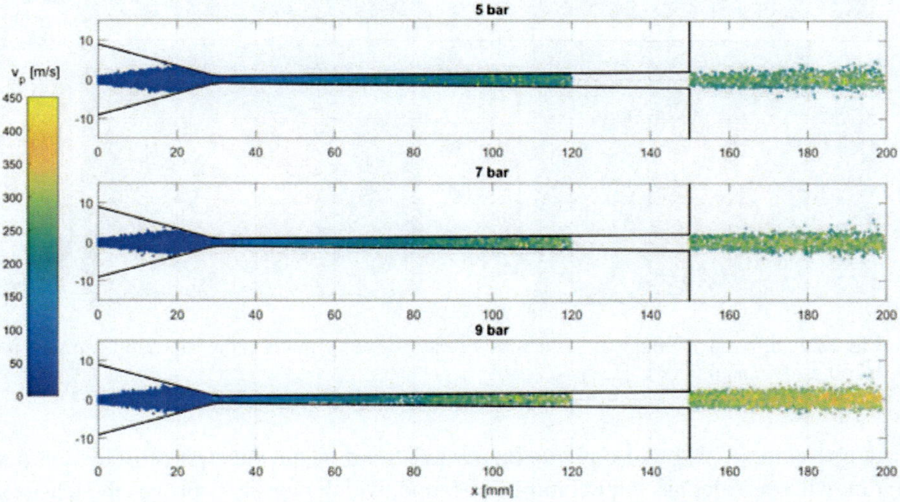

Fig. 2.19 Spatial evolution of particle velocity for Stellite-21 with increasing pressure within nozzle [86]

2.3.3 Substrate and Coating Temperature Measurement

Experimental measurement on the substrate surface temperature during the preheating process was carried out in recent years [36, 59, 116, 131, 138]. Basically, two tools can be used to realize the measurement, namely, thermocouple and infrared camera. Thermocouple as the most direct and common way was used for measuring Al [59] and Ti [36] substrates in previous works. In those experiments, thermocouples were inserted into small holes from the substrate back surface; the holes were drilled sufficiently deep to assure the thermocouple was as close as possible to the substrate front surface. The measured data showed that temperature reduced from the stagnation point towards the surrounding, which was in agreement with the CFD modeling results.

Alternatively, infrared camera as another option was also used in some works [116, 131]. The advantage of this technique is that the general temperature distribution on the substrate surface can be recorded rather than the single point temperature. An instance is shown in Fig. 2.20 which shows the infrared photographs of the substrate surface temperature after heating by the main gas with different inlet

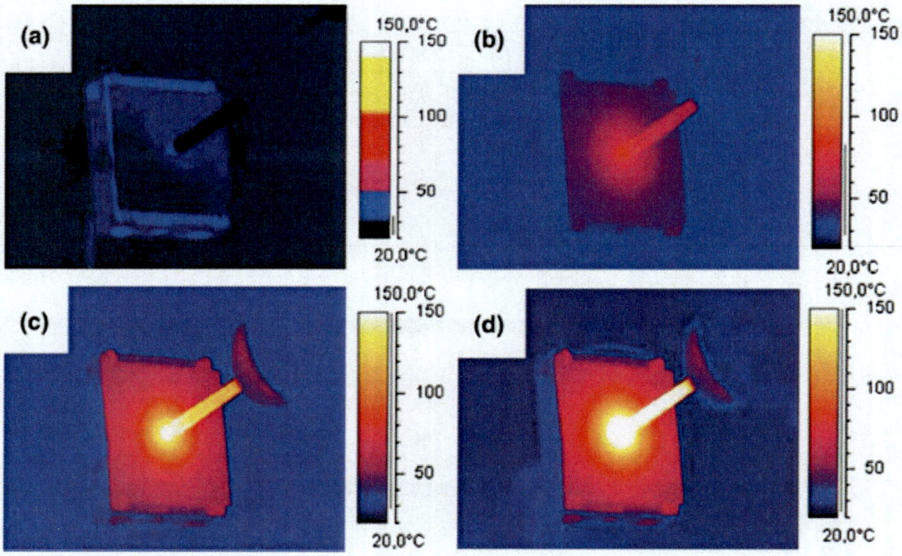

Fig. 2.20 Infrared photographs of the substrate surface temperature at the inlet temperature of **a** 33, **b** 200, **c** 400 and **d** 500 °C [131]

temperatures [131]. An obvious increasing trend of the substrate surface temperature with the inlet gas temperature was found, which positively proves the feasibility of infrared technique. Studies based on infrared camera measurement also revealed that Nusselt number on the substrate surface increased with the increment of total pressure but insensitive to total temperature; heat transfer coefficient increased as the total temperature and surface roughness were increased but independent on the substrate thickness [138]. In addition, infrared camera also allows measuring the coating temperature during the deposition process because it is not necessary to contact with the measured target as thermocouple does [139]. In summary, experimental measurement of substrate surface and coating temperature is not quite popular so far. In most cases, it was only used for validating the CFD modeling results. But its potential application on the coating temperature measurement may attract more and more interests in the near future as the temperature is a very important concern of the high-quality cold sprayed coating.

References

1. Grujicic, M., Tong, C., DeRosset, W.S., Helfritch, D.: Flow analysis and nozzle-shape optimization for the cold-gas dynamic-spray process. Proc. Inst. Mech. Eng. Part B J. Eng. Manuf. **217**, 1603–1613 (2003). https://doi.org/10.1243/095440503771909980
2. Grujicic, M., Zhao, C.L., Tong, C., DeRosset, W.S., Helfritch, D.: Analysis of the impact velocity of powder particles in the cold-gas dynamic-spray process. Mater. Sci. Eng. **368**, 222–230 (2004). https://doi.org/10.1016/j.msea.2003.10.312

3. Jodoin, B., Raletz, F., Vardelle, M.: Cold spray modeling and validation using an optical diagnostic method. Surf. Coatings Technol. **200**, 4424–4432 (2006). https://doi.org/10.1016/j.surfcoat.2005.02.209
4. Pardhasaradhi, S.P., Venkatachalapathy, V., Joshi, S.V., Govindan, S.: Optical diagnostics study of gas particle transport phenomena in cold gas dynamic spraying and comparison with model predictions. J. Therm. Spray Technol. **17**, 551–563 (2008). https://doi.org/10.1007/s11666-008-9206-0
5. Jodoin, B.: Cold spray nozzle mach number limitation. J. Therm. Spray Technol. **11**, 496–507 (2002). https://doi.org/10.1361/105996302770348628
6. Dykhuizen, R.C., Smith, M.F.: Gas dynamic principles of cold spray. J. Therm. Spray Technol. **7**, 205–212 (1998). https://doi.org/10.1361/105996398770350945
7. Stoltenhoff, T., Kreye, H., Richter, H.J.: An analysis of the cold spray process and its coatings. J. Therm. Spray Technol. **11**, 542–550 (2002). https://doi.org/10.1361/105996302770348682
8. Han, T., Zhao, Z., Gillispie, B.A., Smith, J.R.: Effects of spray conditions on coating formation by the kinetic spray process. J. Therm. Spray Technol. **14**, 373–383 (2005). https://doi.org/10.1361/105996305X59369
9. Lee, M.W., Park, J.J., Kim, D.Y., Yoon, S.S., Kim, H.Y., James, S.C., Chandra, S., Coyle, T.: Numerical studies on the effects of stagnation pressure and temperature on supersonic flow characteristics in cold spray applications. J. Therm. Spray Technol. **20**, 1085–1097 (2011). https://doi.org/10.1007/s11666-011-9641-1
10. Champagne, V.K., Helfritch, D.J., Dinavahi, S.P.G., Leyman, P.F.: Theoretical and experimental particle velocity in cold spray. J. Therm. Spray Technol. **20**, 425–431 (2011). https://doi.org/10.1007/s11666-010-9530-z
11. Klinkov, S.V., Kosarev, V.F., Sova, A.A., Smurov, I.: Calculation of particle parameters for cold spraying of metal-ceramic mixtures. J. Therm. Spray Technol. **18**, 944–956 (2009). https://doi.org/10.1007/s11666-009-9346-x
12. Winnicki, M., Małachowska, A., Dudzik, G., Rutkowska-Gorczyca, M., Marciniak, M., Abramski, K., Ambroziak, A., Pawłowski, L.: Numerical and experimental analysis of copper particles velocity in low-pressure cold spraying process. Surf. Coatings Technol. **268**, 230–240 (2015). https://doi.org/10.1016/j.surfcoat.2014.11.059
13. Raletz, F.F., Vardelle, M., Ezoo, G.: Critical particle velocity under cold spray conditions. Surf. Coatings Technol. **201**, 1942–1947 (2006). https://doi.org/10.1016/j.surfcoat.2006.04.061
14. Yin, S., Zhang, M., Guo, Z., Liao, H., Wang, X.: Numerical investigations on the effect of total pressure and nozzle divergent length on the flow character and particle impact velocity in cold spraying. Surf. Coatings Technol. **232**, 290–297 (2013). https://doi.org/10.1016/j.surfcoat.2013.05.017
15. Alhulaifi, A.S., Buck, G.A.: Characterization of the interacting gas-particle dynamic system in cold spray nozzles in dimensional and non-dimensional form. Powder Technol. **338**, 325–332 (2018). https://doi.org/10.1016/j.powtec.2018.07.032
16. Ozdemir, O.C., Widener, C.A., Carter, M.J., Johnson, K.W.: Predicting the effects of powder feeding rates on particle impact conditions and cold spray deposited coatings. J. Therm. Spray Technol. **26**, 1598–1615 (2017). https://doi.org/10.1007/s11666-017-0611-0
17. Alkhimov, A.P., Kosarev, V.F., Klinkov, S.V.: The features of cold spray nozzle dessign. J. Therm. Spray Technol. **10**, 375–381 (2001). https://doi.org/10.1361/105996301770349466
18. Kosarev, V.F., Klinkov, S.V., Alkhimov, A.P., Papyrin, A.N.: On some aspects of gas dynamics of the cold spray process. J. Therm. Spray Technol. **12**, 265–281 (2003). https://doi.org/10.1361/105996303770348384
19. Assadi, H., Schmidt, T., Richter, H., Kliemann, J.O., Binder, K., Gärtner, F., Klassen, T., Kreye, H.: On parameter selection in cold spraying. J. Therm. Spray Technol. **20**, 1161–1176 (2011). https://doi.org/10.1007/s11666-011-9662-9
20. Versteeg, H., Malalasekera, W.: An introduction to computational fluid dynamics: the finite volume method. Pearson Education Limited (2007)
21. Samareh, B., Stier, O., Lüthen, V., Dolatabadi, A.: Assessment of CFD modeling via Flow visualization in cold spray process. J. Therm. Spray Technol. **18**, 934–943 (2009). https://doi.org/10.1007/s11666-009-9363-9

22. Han, T., Gillispie, B.A., Zhao, Z.B.: An investigation on powder injection in the high-pressure cold spray process. J. Therm. Spray Technol. **18**, 320–330 (2009). https://doi.org/10.1007/s11666-009-9329-y
23. Farvardin, E., Stier, O., Lüthen, V., Dolatabadi, A.: Effect of using liquid feedstock in a high pressure cold spray nozzle. J. Therm. Spray Technol. **20**, 307–316 (2011). https://doi.org/10.1007/s11666-010-9597-6
24. Taylor, K., Jodoin, B., Karov, J.: Particle loading effect in cold spray. J. Therm. Spray Technol. **15**, 273–279 (2006). https://doi.org/10.1361/105996306X108237
25. Tabbara, H., Gu, S., McCartney, D.G., Price, T.S., Shipway, P.H.: Study on process optimization of cold gas spraying. J. Therm. Spray Technol. **20**, 608–620 (2011). https://doi.org/10.1007/s11666-010-9564-2
26. Huang, G., Gu, D., Li, X., Xing, L., Wang, H.: Numerical simulation on syphonage effect of laval nozzle for low pressure cold spray system. J. Mater. Process. Technol. **214**, 2497–2504 (2014). https://doi.org/10.1016/j.jmatprotec.2014.05.014
27. Caruso, F., Meyer, M.C., Lupoi, R.: Three-dimensional numerical simulations of the particle loading effect on gas flow features for low pressure cold spray applications. Surf. Coatings Technol. **339**, 181–190 (2018). https://doi.org/10.1016/j.surfcoat.2018.02.016
28. GHIAS, R.: Simulation of flow through supersonic cruise nozzle:a validaiton study. Presented at the (2011)
29. Pattison, J., Celotto, S., Khan, A., O'Neill, W.: Standoff distance and bow shock phenomena in the cold spray process. Surf. Coatings Technol. **202**, 1443–1454 (2008). https://doi.org/10.1016/j.surfcoat.2007.06.065
30. Karimi, M., Fartaj, A., Rankin, G., Vanderzwet, D., Birtch, W., Villafuerte, J.: Numerical simulation of the cold gas dynamic spray process. J. Therm. Spray Technol. **15**, 518–523 (2006). https://doi.org/10.1361/105996306X146866
31. Samareh, B., Dolatabadi, A.: A three-dimensional analysis of the cold spray process: The effects of substrate location and shape. J. Therm. Spray Technol. **16**, 634–642 (2007). https://doi.org/10.1007/s11666-007-9082-z
32. Sova, A., Doubenskaia, M., Grigoriev, S., Okunkova, A., Smurov, I.: Parameters of the gas-powder supersonic jet in cold spraying using a mask. J. Therm. Spray Technol. **22**, 551–556 (2013). https://doi.org/10.1007/s11666-013-9891-1
33. Breuninger, P., Krull, F., Huttenlochner, K., Müller-Reno, C., Ziegler, C., Merz, R., Kopnarski, M., Antonyuk, S.: Microstructuring of steel surfaces via cold spraying with 316L particles for studying the particle-wall collision behavior. Surf. Coatings Technol. **379**, 125054 (2019). https://doi.org/10.1016/j.surfcoat.2019.125054
34. Raoelison, R.N., Koithara, L.L., Costil, S., Langlade, C.: Turbulences of the supersonic gas flow during cold spraying and their negative effects: A DNS CFD analysis coupled with experimental observation and laser impulse high-speed shadowgraphs of the particles in-flight flow. Int. J. Heat Mass Transf. **147**, 118894 (2020). https://doi.org/10.1016/j.ijheatmasstransfer.2019.118894
35. Li, C., Singh, N., Andrews, A., Olson, B.A., Schwartzentruber, T.E., Hogan, C.J.: Mass, momentum, and energy transfer in supersonic aerosol deposition processes. Int. J. Heat Mass Transf. **129**, 1161–1171 (2019). https://doi.org/10.1016/j.ijheatmasstransfer.2018.10.028
36. Zahiri, S.H., Phan, T.D., Masood, S.H., Jahedi, M.: Development of holistic three-dimensional models for cold spray supersonic jet. J. Therm. Spray Technol. **23**, 919–933 (2014). https://doi.org/10.1007/s11666-014-0113-2
37. Faizan-Ur-Rab, M., Zahiri, S.H.H., Masood, S.H.H., Phan, T.D.D., Jahedi, M., Nagarajah, R.: Application of a holistic 3D model to estimate state of cold spray titanium particles. Mater. Des. **89**, 1227–1241 (2015). https://doi.org/10.1016/j.matdes.2015.10.075
38. Hadjadj, A., Perrot, Y., Verma, S.: Numerical study of shock/boundary layer interaction in supersonic overexpanded nozzles. Aerosp. Sci. Technol. **42**, 158–168 (2015). https://doi.org/10.1016/j.ast.2015.01.010
39. Balabel, A., Hegab, A.M., Nasr, M., El-Behery, S.M.: Assessment of turbulence modeling for gas flow in two-dimensional convergent-divergent rocket nozzle. Appl. Math. Model. **35**, 3408–3422 (2011). https://doi.org/10.1016/j.apm.2011.01.013

40. Nasif, G., Barron, R.M., Balachandar, R., Villafuerte, J.: Numerical assessment of miniaturized cold spray nozzle for additive manufacturing. Int. J. Numer. Methods Heat Fluid Flow. **29**, 2277–2296 (2019). https://doi.org/10.1108/HFF-10-2018-0553
41. Liebersbach, P., Foelsche, A., Champagne, V.K., Siopis, M., Nardi, A., Schmidt, D.P.: CFD simulations of feeder tube pressure oscillations and prediction of clogging in cold spray nozzles. J. Therm. Spray Technol. **29**, 400–412 (2020). https://doi.org/10.1007/s11666-020-00992-0
42. Zapryagaev, V.I., Kudryavtsev, A.N., Lokotko, A.V., Solotchin, A.V., Pavlov, A.A., Hadjadj, A.: An experimental and numberical study of a supersonic-jet shock-wave structure. In: Proceedings of the XI International Conference on the Methods of Aerophysical Research. pp. 187–191, Russia (2002)
43. Alhulaifi, A.S., Buck, G.A.: A simplified approach for the determination of critical velocity for cold spray processes. J. Therm. Spray Technol. **23**, 1259–1269 (2014). https://doi.org/10.1007/s11666-014-0128-8
44. Katanoda, H., Fukuhara, M., Iino, N.: Numerical study of combination parameters for particle impact velocity and temperature in cold spray. J. Therm. Spray Technol. **16**, 627–633 (2007). https://doi.org/10.1007/s11666-007-9087-7
45. Yin, S., Wang, X.F., Li, W.Y., Xu, B.P.: Numerical study on the effect of substrate angle on particle impact velocity and normal velocity component in cold gas dynamic spraying based on CFD. J. Therm. Spray Technol. **19**, 1155–1162 (2010). https://doi.org/10.1007/s11666-010-9510-3
46. Sova, A., Grigoriev, S., Kochetkova, A., Smurov, I.: Influence of powder injection point position on efficiency of powder preheating in cold spray: Numerical study. Surf. Coatings Technol. **242**, 226–231 (2014). https://doi.org/10.1016/j.surfcoat.2013.10.078
47. Li, W.Y., Liao, H., Wang, H.T., Li, C.J., Zhang, G., Coddet, C.: Optimal design of a convergent-barrel cold spray nozzle by numerical method. Appl. Surf. Sci. **253**, 708–713 (2006). https://doi.org/10.1016/j.apsusc.2005.12.157
48. Li, W.Y., Li, C.J.: Optimal design of a novel cold spray gun nozzle at a limited space. J. Therm. Spray Technol. **14**, 391–396 (2005). https://doi.org/10.1361/105996305X59404
49. Li, W.-Y., Liao, H., Douchy, G., Coddet, C.: Optimal design of a cold spray nozzle by numerical analysis of particle velocity and experimental validation with 316L stainless steel powder. Mater. Des. **28**, 2129–2137 (2007). https://doi.org/10.1016/j.matdes.2006.05.016
50. Kwon, E.H., Cho, S.H., Han, J.W., Lee, C.H., Kim, H.J.: Particle Behavior in Supersonic Flow during the Cold Spray Process. Matals Mater. Int. **11**, 377–381 (2005). https://doi.org/10.1007/BF03027508
51. Yin, S., Wang, X.F., Li, W.Y.: Computational analysis of the effect of nozzle cross-section shape on gas flow and particle acceleration in cold spraying. Surf. Coatings Technol. **205**, 2970–2977 (2011). https://doi.org/10.1016/j.surfcoat.2010.11.002
52. Suo, X.K., Liu, T.K., Li, W.Y., Suo, Q.L., Planche, M.P., Liao, H.L.: Numerical study on the effect of nozzle dimension on particle distribution in cold spraying. Surf. Coatings Technol. **220**, 107–111 (2013). https://doi.org/10.1016/j.surfcoat.2012.09.029
53. Li, W.Y., Yin, S., Guo, X., Liao, H., Wang, X.F., Coddet, C.: An investigation on temperature distribution within the substrate and nozzle wall in cold spraying by numerical and experimental methods. J. Therm. Spray Technol. **21**, 41–48 (2012). https://doi.org/10.1007/s11666-011-9685-2
54. Yin, S., Wang, X.F., Li, W.Y., Li, Y.: Numerical study on the effect of substrate size on the supersonic jet flow and temperature distribution within the substrate in cold spraying (2012)
55. Alhulaifi, A.S., Buck, G.A., Arbegast, W.J.: Numerical and experimental investigation of cold spray gas dynamic effects for polymer coating. J. Therm. Spray Technol. **21**, 852–862 (2012). https://doi.org/10.1007/s11666-012-9743-4
56. Li, W.Y., Zhang, C., Guo, X.P., Zhang, G., Liao, H.L., Li, C.J., Coddet, C.: Effect of standoff distance on coating deposition characteristics in cold spraying. Mater. Des. **29**, 297–304 (2008). https://doi.org/10.1016/j.matdes.2007.02.005

57. Li, W.Y., Li, C.J.: Optimization of spray conditions in cold spraying based on numerical analysis of particle velocity. Trans. Nonferrous Met. Soc. China (English Ed.). **14**, 43–48 (2004)
58. Faizan-Ur-Rab, M., Zahiri, S.H., Masood, S.H., Jahedi, M., Nagarajah, R.: PIV validation of 3D multicomponent model for cold spray within nitrogen and helium supersonic flow field. J. Therm. Spray Technol. **26**, 941–957 (2017). https://doi.org/10.1007/s11666-017-0567-0
59. Yin, S., Wang, X.F., Li, W.Y., Guo, X.P.: Examination on substrate preheating process in cold gas dynamic spraying. J. Therm. Spray Technol. **20**, 852–859 (2011). https://doi.org/10.1007/s11666-011-9623-3
60. Wang, X.-F., Yin, S., Xu, B.P.B.P.B.P.: Effect of cold spray particle conditions and optimal standoff distance on impact velocity. J. Dalian Univ. Technol. **51**, 498–504 (2011)
61. Harihara Sudhan, K., Krishna Prasad, G., Kothurkar, N.K., Srikrishnan, A.R.: Studies on supersonic cold spray deposition of microparticles using a bell-type nozzle. Surf. Coatings Technol. **383**, 125244 (2020). https://doi.org/10.1016/j.surfcoat.2019.125244
62. Meyer, M., Lupoi, R.: An analysis of the particulate flow in cold spray nozzles. Mech. Sci. **6**, 127–136 (2015). https://doi.org/10.5194/ms-6-127-2015
63. Jung, H.B., Park, J.I., Park, S.H., Kim, H.J., Lee, C.H., Han, J.W.: Effect of the expansion ratio and length ratio on a gas-particle flow in a converging-diverging cold spray nozzle. Met. Mater. Int. **15**, 967–970 (2009). https://doi.org/10.1007/s12540-009-0967-x
64. Wang, X., Zhang, B., Lv, J., Yin, S.: Investigation on the clogging behavior and additional wall cooling for the axial-injection cold spray nozzle. J. Therm. Spray Technol. **24**, 696–701 (2015). https://doi.org/10.1007/s11666-015-0227-1
65. Tang, W., Liu, J., Chen, Q., Zhang, X., Chen, Z.: The effects of two gas flow streams with initial temperature and pressure differences in cold spraying nozzle. Surf. Coatings Technol. **240**, 86–95 (2014). https://doi.org/10.1016/j.surfcoat.2013.12.019
66. Jen, T.C., Li, L., Cui, W., Chen, Q., Zhang, X.: Numerical investigations on cold gas dynamic spray process with nano—and microsize particles. Int. J. Heat Mass Transf. **48**, 4384–4396 (2005). https://doi.org/10.1016/j.ijheatmasstransfer.2005.05.008
67. Ning, X.J., Wang, Q.S., Ma, Z., Kim, H.J.: Numerical study of in-flight particle parameters in low-pressure cold spray process. J. Therm. Spray Technol. **19**, 1211–1217 (2010). https://doi.org/10.1007/s11666-010-9548-2
68. Yin, S., Liu, Q., Liao, H., Wang, X.: Effect of injection pressure on particle acceleration, dispersion and deposition in cold spray. Comput. Mater. Sci. **90**, 7–15 (2014). https://doi.org/10.1016/j.commatsci.2014.03.055
69. Park, J.J., Lee, M.W., Yoon, S.S., Kim, H.Y., James, S.C., Heister, S.D., Chandra, S., Yoon, W.H., Park, D.S., Ryu, J.: Supersonic nozzle flow simulations for particle coating applications: effects of shockwaves, nozzle geometry, ambient pressure, and substrate location upon flow characteristics. J. Therm. Spray Technol. **20**, 514–522 (2011). https://doi.org/10.1007/s11666-010-9542-8
70. Yin, S., Suo, X., Liao, H., Guo, Z., Wang, X.: Significant influence of carrier gas temperature during the cold spray process. Surf. Eng. **30**, 443–450 (2014). https://doi.org/10.1179/1743294414y.0000000276
71. Suo, X., Yin, S., Planche, M.P., Liu, T., Liao, H.: Strong effect of carrier gas species on particle velocity during cold spray processes. Surf. Coatings Technol. **268**, 90–93 (2015). https://doi.org/10.1016/j.surfcoat.2014.04.039
72. Jen, T.C., Pan, L., Li, L., Chen, Q., Cui, W.: The acceleration of charged nano-particles in gas stream of supersonic De-Lavel-type nozzle coupled with static electric field. Appl. Therm. Eng. **27**, 2877–2885 (2007). https://doi.org/10.1016/j.applthermaleng.2005.07.030
73. Huang, G., Gu, D., Li, X., Xing, L.: Computational simulation on a coaxial substream powder feeding laval nozzle of cold spraying. Mater. Sci. **20** (2014)
74. Huang, R., Fukanuma, H.: Study of the influence of particle velocity on adhesive strength of cold spray deposits. J. Therm. Spray Technol. **21**, 541–549 (2012). https://doi.org/10.1007/s11666-011-9707-0

75. Lupoi, R., O'Neill, W.: Powder stream characteristics in cold spray nozzles. Surf. Coatings Technol. **206**, 1069–1076 (2011). https://doi.org/10.1016/j.surfcoat.2011.07.061
76. Zhu, W., Zhang, X., Zhang, M., Tian, X., Li, D.: Integral numerical modeling of the deposition profile of a cold spraying process as an additive manufacturing technology. Prog. Addit. Manuf. **4**, 357–370 (2019). https://doi.org/10.1007/s40964-018-0071-1
77. Garmeh, S., Jadidi, M., Dolatabadi, A.: Three-dimensional modeling of cold spray for additive manufacturing. J. Therm. Spray Technol. **29**, 38–50 (2020). https://doi.org/10.1007/s11666-019-00928-3
78. Chun, D.-M.M., Choi, J.-O.O., Lee, C.S., Ahn, S.-H.H.: Effect of stand-off distance for cold gas spraying of fine ceramic particles (<5μm) under low vacuum and room temperature using nano-particle deposition system (NPDS). Surf. Coatings Technol. **206**, 2125–2132 (2012). https://doi.org/10.1016/j.surfcoat.2011.09.043
79. Ozdemir, O.C., Chen, Q., Muftu, S., Champagne, V.K.: Modeling the continuous heat generation in the cold spray coating process. J. Therm. Spray Technol. **28**, 108–123 (2019). https://doi.org/10.1007/s11666-018-0794-z
80. Sova, A., Okunkova, A., Grigoriev, S., Smurov, I.: Velocity of the particles accelerated by a cold spray micronozzle: experimental measurements and numerical simulation. J. Therm. Spray Technol. **22**, 75–80 (2013). https://doi.org/10.1007/s11666-012-9846-y
81. Karimi, M., Rankin, G.W., Fartaj, A.: Parametric study of exhaust pattern in cold spray using CFD and particle-wall impact analysis. J. Appl. Fluid Mech. **7**, 75–87 (2014)
82. Samareh, B., Dolatabadi, A.: Dense Particulate Flow in a Cold Gas Dynamic Spray System. J. Fluids Eng. **130**, 081702 (2008). https://doi.org/10.1115/1.2957914
83. Takana, H., Ogawa, K., Shoji, T., Nishiyama, H.: Computational simulation of cold spray process assisted by electrostatic force. Powder Technol. **185**, 116–123 (2008). https://doi.org/10.1016/j.powtec.2007.10.005
84. Li, S., Muddle, B., Jahedi, M., Soria, J.: A numerical investigation of the cold spray process using underexpanded and overexpanded jets. J. Therm. Spray Technol. **21**, 108–120 (2012). https://doi.org/10.1007/s11666-011-9691-4
85. Meyer, M., Yin, S., Lupoi, R.: Particle in-flight velocity and dispersion measurements at increasing particle feed rates in cold spray. J. Therm. Spray Technol. **26**, 60–70 (2017). https://doi.org/10.1007/s11666-016-0496-3
86. Meyer, M., Caruso, F., Lupoi, R.: Particle velocity and dispersion of high Stokes number particles by PTV measurements inside a transparent supersonic Cold Spray nozzle. Int. J. Multiph. Flow. **106**, 296–310 (2018). https://doi.org/10.1016/j.ijmultiphaseflow.2018.05.018
87. Clift, R., Grace, J.R., Weber, M.E.: Bubbles, Drops, and Particles. In: Press, A. (ed.) Technical Report (1978)
88. Crowe, C., Sommerfeld, M., Tsuji, Y.: Multiphase flows with droplets and particles. CRC Press, Boca Raton (1998)
89. Leitz, K.H., O'Sullivan, M., Plankensteiner, A., Kestler, H., Sigl, L.S.: OpenFOAM modeling of particle heating and acceleration in cold spraying. J. Therm. Spray Technol. **27**, 135–144 (2018). https://doi.org/10.1007/s11666-017-0644-4
90. Leitz, K.H., O'Sullivan, M., Plankensteiner, A., Lichtenegger, T., Pirker, S., Kestler, H., Sigl, L.S.: CFDEM modelling of particle heating and acceleration in cold spraying. Int. J. Refract. Met. Hard Mater. **73**, 192–198 (2018). https://doi.org/10.1016/j.ijrmhm.2018.02.003
91. Schiller, L., N.: A drag coefficient correlation. VDI Zeits. **77**, 318–320 (1933)
92. Morsi, S.A., Alexander, A.J.: An investigation of particle trajectories in two-phase flow systems. J. Fluid Mech. **55**, 193–208 (1972)
93. Haider, A., Levenspiel, O.: Drag coefficient and terminal velocity of spherical and nonspherical particles. **58**, 63–70 (1989)
94. Crowe, C.T.: Drag coefficient of particles in a rocket nozzle. AIAA J. **5**, 1021–1022 (1967)
95. Henderson, C.B.C.B.: Drag coefficients of spheres in continuum and rarefied flows. AIAA J. **14**, 707–708 (1976). https://doi.org/10.2514/3.61409
96. Katanoda, H.: Numerical simulation of temperature uniformity within solid particles in cold spray. J. Solid Mech. Mater. Eng. **2**, 58–69 (2008). https://doi.org/10.1299/jmmp.2.58

97. Dewar, M.P., McDonald, A.G., Gerlich, A.P.: Interfacial heating during low-pressure cold-gas dynamic spraying of aluminum coatings. J. Mater. Sci. **47**, 184–198 (2012). https://doi.org/10.1007/s10853-011-5786-z
98. Ranz, W.E., Jr, W.R.M.: Evaporation from drops Part I. Chem. Eng. Prog. 48, 141–146 (1952)
99. Hughmark, G.A.: Mass and heat transfer from rigid sphere. AlChE J. **13**, 1219–1221 (1967)
100. Carlson, D.J., Chambre, P.L.: Particle drag and heat transfer in rocket nozzles. AIAA J. **2**, 1980–1984 (1964)
101. Y., P.W., Prasad, V., Wang, G.-X., Sampath, S., Fincke, J.R.: Model and powder particle heating, melting, resolidification, and evaporation in plasma spraying processes. J. Heat Transfer. **121**, 691–699 (1999)
102. K.W., T.: Time dependent boundary conditions for hyperbolic systems. J. Comput. Phys. **68**, 1–24 (1987)
103. Gärtner, F., Stoltenhoff, T., Schmidt, T., Kreye, H.: The cold spray process and its potential for industrial applications. J. Therm. Spray Technol. **15**, 223–232 (2006). https://doi.org/10.1361/105996306X108110
104. Schmidt, T., Assadi, H., Gärtner, F., Richter, H., Stoltenhoff, T., Kreye, H., Klassen, T.: From particle acceleration to impact and bonding in cold spraying. J. Therm. Spray Technol. **18**, 794–808 (2009). https://doi.org/10.1007/s11666-009-9357-7
105. Schmidt, T., Gaertner, F., Kreye, H.: New developments in cold spray based on higher gas and particle temperatures. J. Therm. Spray Technol. **15**, 488–494 (2006). https://doi.org/10.1361/105996306X147144
106. Gilmore, D.L.L., Dykhuizen, R.C.C., Neiser, R.A.A., Roemer, T.J.J., Smith, M.F.F.: Particle velocity and deposition efficiency in the cold spray process. J. Therm. Spray Technol. **8**, 576–582 (1999). https://doi.org/10.1361/105996399770350278
107. Chavan, N.M., Vinod Kumar, M., Sudharshan Phani, P., Pant, P., Sundararajan, G.: Influence of nozzle throat cross section on microstructure and properties of cold sprayed coatings. J. Therm. Spray Technol. **28**, 1718–1729 (2019). https://doi.org/10.1007/s11666-019-00913-w
108. Kiselev, S.P., Kiselev, V.P., Klinkov, S.V., Kosarev, V.F., Zaikovskii, V.N.: Study of the gas-particle radial supersonic jet in the cold spraying. Surf. Coatings Technol. **313**, 24–30 (2017). https://doi.org/10.1016/j.surfcoat.2017.01.046
109. Nickel, R., Bobzin, K., Lugscheider, E., Parkot, D., Varava, W., Olivier, H., Luo, X.: Numerical studies of the application of shock tube technology for cold gas dynamic spray process. J. Therm. Spray Technol. **16**, 729–735 (2007). https://doi.org/10.1007/s11666-007-9123-7
110. Ozdemir, O.C., Widener, C.A.: Influence of powder injection parameters in high-pressure cold spray. J. Therm. Spray Technol. **26**, 1411–1422 (2017). https://doi.org/10.1007/s11666-017-0606-x
111. Schmidt, T., Gärtner, F., Assadi, H., Kreye, H.: Development of a generalized parameter window for cold spray deposition. Acta Mater. **54**, 729–742 (2006). https://doi.org/10.1016/j.actamat.2005.10.005
112. Yin, S., Suo, X., Guo, Z., Liao, H., Wang, X.: Deposition features of cold sprayed copper particles on preheated substrate. Surf. Coatings Technol. **268**, 252–256 (2015). https://doi.org/10.1016/j.surfcoat.2014.11.009
113. Fukumoto, M., Wada, H., Tanabe, K., Yamada, M., Yamaguchi, E., Niwa, A., Sugimoto, M., Izawa, M.: Effect of substrate temperature on deposition behavior of copper particles on substrate surfaces in the cold spray process. J. Therm. Spray Technol. **16**, 643–650 (2007). https://doi.org/10.1007/s11666-007-9121-9
114. Yin, S., Suo, X., Xie, Y., Li, W., Lupoi, R., Liao, H.: Effect of substrate temperature on interfacial bonding for cold spray of Ni onto Cu. J. Mater. Sci. **50**, 7448–7457 (2015). https://doi.org/10.1007/s10853-015-9304-6
115. Yin, S., Sun, Y., Wang, X., Guo, Z., Liao, H.: Effect of spray angle on temperature distribution within the metallic substrate in cold spraying. J. Therm. Spray Technol. **22**, 983–991 (2013). https://doi.org/10.1007/s11666-013-9931-x
116. McDonald, A.G., Ryabinin, A.N.N., Irissou, E., Legoux, J.-G.G., McDonald, A.G., Legoux, J.-G.G.: Simulation of gas-substrate heat exchange during cold-gas dynamic spraying. Int. J. Therm. Sci. **56**, 12–18 (2012). https://doi.org/10.1016/j.ijthermalsci.2012.01.007

117. Chen, C., Xie, Y., Verdy, C., Huang, R., Liao, H., Ren, Z., Deng, S.: Numerical investigation of transient coating build-up and heat transfer in cold spray. Surf. Coatings Technol. **326**, 355–365 (2017). https://doi.org/10.1016/j.surfcoat.2017.07.069
118. Klinkov, S.V., Kosarev, V.F., Zaikovskii, V.N.: Influence of flow swirling and exit shape of barrel nozzle on cold spraying. J. Therm. Spray Technol. **20**, 837–844 (2011). https://doi.org/10.1007/s11666-011-9621-5
119. Wong, W., Irissou, E., Ryabinin, A.N., Legoux, J.G., Yue, S.: Influence of helium and nitrogen gases on the properties of cold gas dynamic sprayed pure titanium coatings. J. Therm. Spray Technol. **20**, 213–226 (2011). https://doi.org/10.1007/s11666-010-9568-y
120. Katanoda, H., Matsuoka, T., Matsuo, K.: Experimental study on shock wave structures in constant-area passage of cold spray nozzle. J. Therm. Sci. **16**, 40–45 (2007). https://doi.org/10.1007/s11630-007-0040-3
121. Smith, M.F., O 'hern, T.J., Brockmann, J.E., Neiser, R.A., Roemer, T.J.: A comparison of two laser-based diagnostics for analysis of particles in thermal spray streams. Proc. 1995 Natl. Therm. Spray Conf. 1–15 (1995)
122. Zahiri, S.H., Yang, W., Jahedi, M.: Characterization of cold spray titanium supersonic jet. J. Therm. Spray Technol. **18**, 110–117 (2009). https://doi.org/10.1007/s11666-008-9278-x
123. Koivuluoto, H., Matikainen, V., Larjo, J., Vuoristo, P.: Novel online diagnostic analysis for in-flight particle properties in cold spraying. J. Therm. Spray Technol. **27**, 423–432 (2018). https://doi.org/10.1007/s11666-018-0685-3
124. Sova, A., Doubenskaia, M., Petrovskiy, P., Smurov, I.: Visualization of particle jet in cold spray by infrared camera: feasibility tests. Int. J. Adv. Manuf. Technol. **95**, 3057–3063 (2018). https://doi.org/10.1007/s00170-017-1435-2
125. Koithara, L.L., Raoelison, R.N., Costil, S.: Flow phenomenon of micron-sized particles during cold spray additive manufacturing: High-speed optic observation and characterization. Adv. Powder Technol. **31**, 1060–1079 (2020). https://doi.org/10.1016/j.apt.2019.12.037
126. Zahiri, S.H., Antonio, C.I., Jahedi, M.: Elimination of porosity in directly fabricated titanium via cold gas dynamic spraying. J. Mater. Process. Technol. **209**, 922–929 (2009). https://doi.org/10.1016/j.jmatprotec.2008.03.005
127. List, A., Gärtner, F., Schmidt, T., Klassen, T.: Impact conditions for cold spraying of hard metallic glasses. J. Therm. Spray Technol. **21**, 531–540 (2012). https://doi.org/10.1007/s11666-012-9750-5
128. Bray, M., Cockburn, A., O'Neill, W.: The Laser-assisted cold spray process and deposit characterisation. Surf. Coatings Technol. **203**, 2851–2857 (2009). https://doi.org/10.1016/j.surfcoat.2009.02.135
129. Champagne, V.K., Helfritch, D., Leyman, P., Grendahl, S., Klotz, B.: Interface material mixing formed by the deposition of copper on aluminum by means of the cold spray process. J. Therm. Spray Technol. **14**, 330–334 (2005). https://doi.org/10.1361/105996305X59332
130. Fukanuma, H., Ohno, N., Sun, B., Huang, R.: In-flight particle velocity measurements with DPV-2000 in cold spray. Surf. Coatings Technol. **201**, 1935–1941 (2006). https://doi.org/10.1016/j.surfcoat.2006.04.035
131. Legoux, J.G., Irissou, E., Moreau, C.: Effect of substrate temperature on the formation mechanism of cold-sprayed aluminum, zinc and tin coatings. J. Therm. Spray Technol. **16**, 619–626 (2007). https://doi.org/10.1007/s11666-007-9091-y
132. Ajdelsztajn, L., Jodoin, B., Richer, P., Sansoucy, E., Lavernia, E.J.: Cold gas dynamic spraying of iron-base amorphous alloy. J. Therm. Spray Technol. **15**, 495–500 (2006). https://doi.org/10.1361/105996306X146857
133. Jodoin, B., Ajdelsztajn, L., Sansoucy, E., Zúñiga, A., Richer, P., Lavernia, E.J.: Effect of particle size, morphology, and hardness on cold gas dynamic sprayed aluminum alloy coatings. Surf. Coatings Technol. **201**, 3422–3429 (2006). https://doi.org/10.1016/j.surfcoat.2006.07.232
134. Mauer, G., Singh, R., Rauwald, K.H., Schrüfer, S., Wilson, S., Vaßen, R.: Diagnostics of cold-sprayed particle velocities approaching critical deposition conditions. J. Therm. Spray Technol. **26**, 1423–1433 (2017). https://doi.org/10.1007/s11666-017-0596-8

135. Wu, J., Fang, H., Yoon, S., Kim, H., Lee, C.: Measurement of particle velocity and characterization of deposition in aluminum alloy kinetic spraying process. Appl. Surf. Sci. **252**, 1368–1377 (2005). https://doi.org/10.1016/j.apsusc.2005.02.108
136. Meyer, M.C., Yin, S., Mcdonnell, K.A., Stier, O., Lupoi, R.: Feed rate effect on particulate acceleration in Cold Spray under low stagnation pressure conditions. Surf. Coatings Technol. in review, 237–245 (2016). https://doi.org/https://doi.org/10.1016/j.surfcoat.2016.07.017
137. Koivuluoto, H., Larjo, J., Marini, D., Pulci, G., Marra, F.: Cold-sprayed Al6061 coatings: online spray. Coatings. **10**, 348 (2020)
138. Mahdavi, A., McDonald, A.: Effect of substrate and process parameters on the gas-substrate convective heat transfer coefficient during cold spraying. J. Therm. Spray Technol. **27**, 433–445 (2018). https://doi.org/10.1007/s11666-017-0660-4
139. Yin, S., Wang, X., Suo, X., Liao, H., Guo, Z., Li, W., Coddet, C.: Deposition behavior of thermally softened copper particles in cold spraying. Acta Mater. **61**, 5105–5118 (2013). https://doi.org/10.1016/j.actamat.2013.04.041

Chapter 3
Manufacturing Parameters for Cold Spray Additive Manufacturing

Abstract Manufacturing parameters are critical for the fabrication of cold sprayed deposits with desirable microstructure and excellent performance. In CSAM, key manufacturing parameters that determine the deposition process include gas pressure, gas temperature, gas type, powder feed rate, nozzle traverse speed, scanning step, standoff distance, spray angle and trajectory. These parameters must work in unison and under careful control to produce high-quality deposits. In this chapter, the aforementioned key manufacturing parameters and their roles in cold spray deposition are introduced and the relationship among these parameters are discussed.

3.1 Summary of the Manufacturing Parameters for CSAM

Figure 3.1 shows a schematic of the typical manufacturing parameters used in CSAM. Manufacturing parameters including the propulsive gas parameters (pressure, temperature and type), powder feeder parameters (powder feed rate), and nozzle parameters (traverse speed, scanning step, standoff distance, spray angle and trajectory) all significantly influence the deposition process, and therefore the deposit quality. Table 3.1 summarizes some typical deposit properties and their relationship with the manufacturing parameters. Understanding these parameters helps the development of the best CSAM manufacturing strategy and products.

3.2 Propulsive Gas Parameters for CSAM

Propulsive gas parameters are the most important manufacturing parameters in CSAM, as they directly determine particle impact velocity and therefore the properties of the deposit. In general, gases used in CSAM include helium, nitrogen and air. The gas pressure and temperature normally range from 0.5 to 6.0 MPa and from 25 to 1000 °C, respectively. It is well understood that higher gas pressure, higher gas temperature or lower gas molecular weight causes higher particle impact velocity, as can be seen in Fig. 3.2 [1, 2]. Using helium instead of nitrogen and air is the most

S. Yin and R. Lupoi, *Cold Spray Additive Manufacturing*, Springer Tracts in Additive Manufacturing, https://doi.org/10.1007/978-3-030-73367-4_3

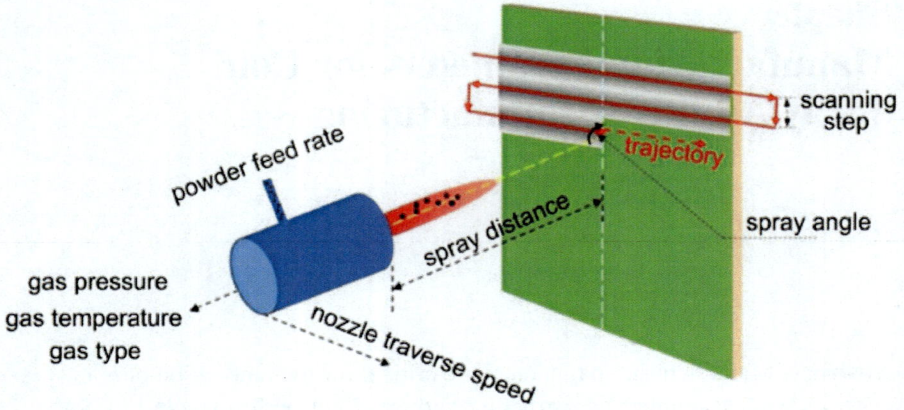

Fig. 3.1 Schematic of the typical manufacturing parameters used in CSAM

Table 3.1 Typical deposit properties affected by the manufacturing parameters

	Porosity	Deposit strength	Adhesion	Residual stress	Deposition efficiency
Gas pressure ↑	↓	↑	↑	↑	↑
Gas temperature ↑	↓	↑	↑	↑	↑
Gas molecular weight ↑	↑	↓	↓	↓	↓
Powder feed rate ↑	↑	↓	↓	↑	↓
Traverse speed ↑	↑	↑	↑	↓	↑
Scanning step	×	×	×	×	×
Standoff distance ↑	○	○	○	○	○
Spray angle ↑	↓	↑	↑	↑	↑
Trajectory	×	×	×	×	×

'↑' increase, '↓' decrease, '○' relevant but no common view, '×' no data available

effective way to increase particle velocity, but the gas cost of helium is also much higher. Otherwise, increasing gas temperature is more effective than increasing gas pressure [1].

To date, it has been widely accepted by the cold spray community that higher particle impact velocity can result in improved deposit properties. As summarized in the Refs. [3, 4], a large number of experimental works demonstrate that increasing particle impact velocity significantly improves deposition efficiency, deposit porosity, deposit strength (cohesion strength or tensile strength) and adhesion strength due to the reduced inherent defects and enhanced metallic bonding.

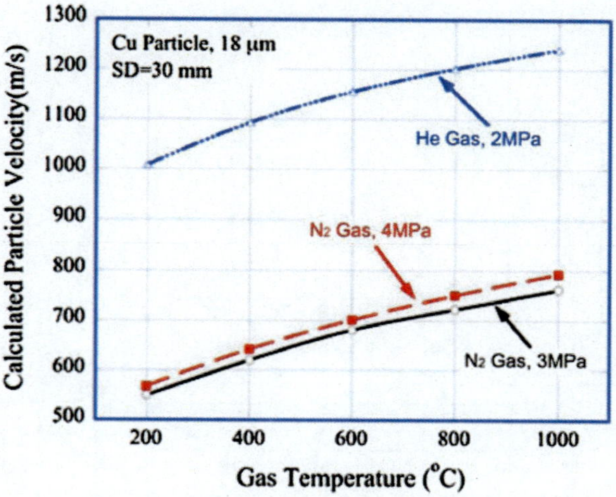

Fig. 3.2 Modeling result of the particle velocity at the location 30 mm away from the nozzle exit as a function of propulsive gas pressure, temperature, and type [1, 2]

Figure 3.3 shows the effect of particle impact velocity (or propulsive gas parameters) on the deposition efficiency, deposit porosity, deposit strength and deposit adhesion strength of cold sprayed titanium deposits [5–7]. Due to the enhanced work-hardening effect, increasing the particle impact velocity will increase the deposit hardness. In addition, studies also suggest that higher impact velocity results in higher residual stress in the cold sprayed deposit due to increased plastic deformation of the particles [8–11]. However, a higher gas temperature was found to relieve the residual stress due to the in-situ annealing effect, therefore meaning that increasing the gas temperature did not significantly increase the residual stresses [12–14].

3.3 Powder Feeder Parameters for CSAM

Powder feed rate is the only parameter controlled by powder feeder. It is defined as the quantity of powder introduced into the nozzle per unit of time. Powder feed rate influences the properties of cold sprayed deposits in three ways. Firstly, powder feed rate significantly affects the flow stream passing through the nozzle and thus the particle velocity [15–18]. Figure 3.4 shows the particle velocity at the nozzle exit as a function of powder mass feed rate [16]. Due to the strong gas-particle interaction at high powder load, particle velocity decreases as the powder feed rate increases. This results in a slight increase in deposit porosity and reduction in deposition efficiency, hardness and tensile strength at increased powder feed rates. However, it should be noted that powder feed rates used in CSAM are typically much lower than 100 g/s, with rates between 10 and 30 g/s being typical. Within this range, the influence of powder feed rate on the particle velocity is not significant [17]. Secondly, powder feed rate also affects the thickness and profile of the single-track deposit; higher powder feed rates result in thicker deposits and sharper track profiles [19, 20]. In

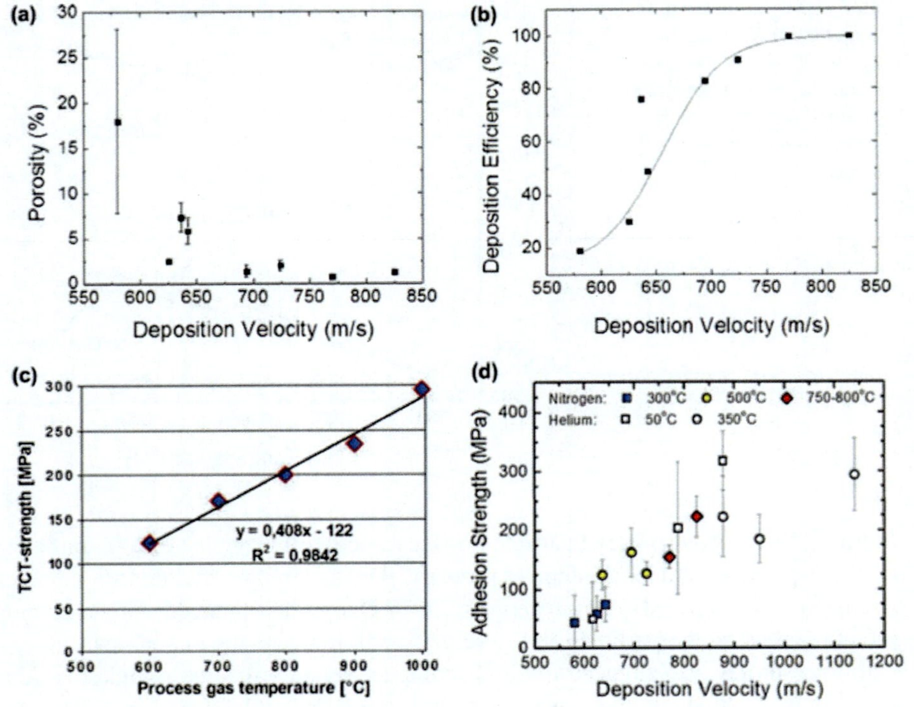

Fig. 3.3 Effect of particle impact velocity (or propulsive gas parameters) on the deposit porosity, deposition efficiency, deposit strength and deposit adhesion strength of cold sprayed titanium deposits. **a**, **b** titanium deposits on mild steel substrate at 2–4 MPa and 300–800 °C with nitrogen [5], **c** titanium deposits on an aluminium substrate at 4 MPa with nitrogen [6], and **d** titanium deposits on a titanium substrate at 1–4 MPa with nitrogen and helium [7]

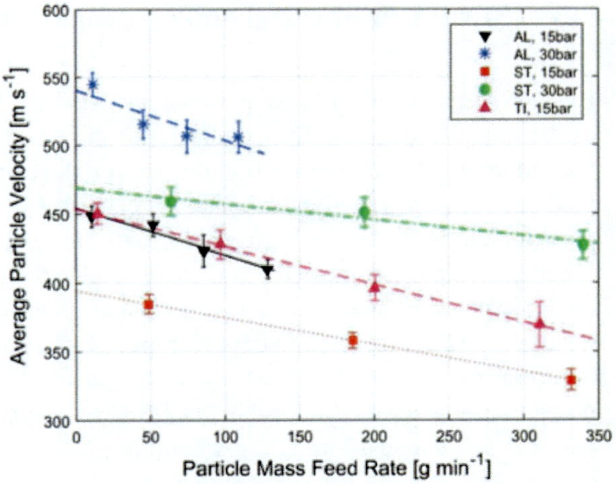

Fig. 3.4 Particle velocity at the nozzle exit as a function of powder mass feed rate [16]

this regard, powder feed rate has a similar effect as nozzle traverse speed. These two parameters must be considered together when determining the profile of a single-track deposit. A detailed description of the effect of these two parameters on the profile of a single-track will be provided in Sect. 3.4. Thirdly, high feed rates can lead to high localized residual stress between deposits and substrate. High residual stresses can cause the delamination of the deposit from the substrate during spraying [17, 18], which should be avoided in CSAM. In summary, the three ways in which powder feed rate affects the deposit properties suggest that powder feed rate should not be too high during the CSAM process.

3.4 Nozzle Traverse Speed

Nozzle traverse speed determines the spray duration and amount of feedstock powder impacting onto the target surface per unit time, thereby directly affecting the deposit thickness and cross-sectional profile of a single-track deposit [17, 19–26]. In general, a lower nozzle traverse speed results in a thicker single-layer deposit and sharper cross-sectional profile, as shown in Fig. 3.5. This can also be achieved through a higher powder feed rate, as addressed in Sect. 3.3 [19, 20]. The variation in peak track thickness due to changing nozzle traverse speed is due to different amount of the feedstock powder released per unit time and per unit area; lower nozzle traverse speed and higher powder feed rate can both result in the release of a greater amount of powder per unit time and per unit area. The change in track profile is caused by the flow characteristics of the propulsive gas as it passes through a supersonic nozzle. According to the fluid dynamics theory, particle velocity and the resultant deposition efficiency at the central zone will be higher than at the outer zone [27–29]. Due to this radial variation in velocity the deposit cross-sectional profile will be a gaussian shape, as shown in Fig. 3.5. The sloped side of the track introduces an angle to the incident particles of any subsequent layers deposited on top of the track, further decreasing the deposition efficiency of the particles and leading to a much steeper sloped side [19, 20]. Therefore, the deposit profile becomes increasingly steeper as the nozzle traverse speed decreases and the powder feed rate increases. In real world applications of CSAM, the control of the thickness and profile of a single-track

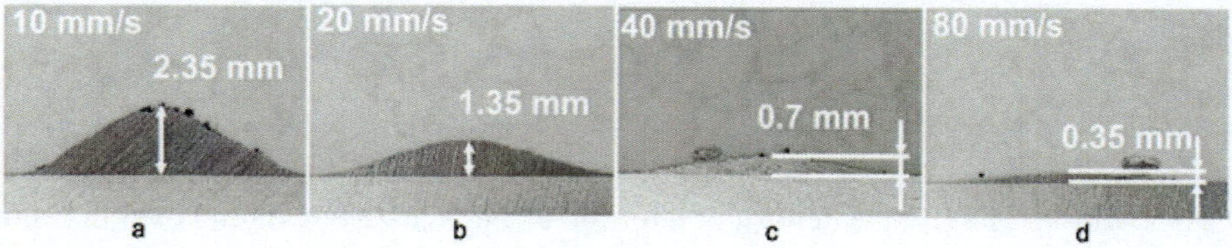

Fig. 3.5 Effect of nozzle traverse speed on the single-track deposit thickness and cross-sectional profile [23]

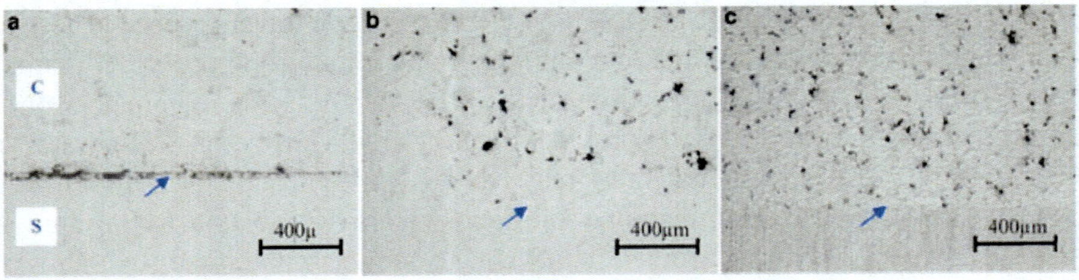

Fig. 3.6 Cross-sectional image of Ti6Al4V deposits onto Ti6Al4V substrates fabricated under different nozzle traverse speed. **a** 100 m/s, **b** 300/s and **c** 500/s. Blue arrows indicate the deposit-substrate interface [24]

deposit is realized by coordination of the nozzle traverse speed and powder feed rate [19, 20].

Nozzle traverse speed also significantly influences the deposit microstructure and properties. The deposit density increases as the nozzle traverse speed decreases, but there is also a reduction in the mechanical properties of the deposit, such as elastic modulus, deposit strength, and adhesion strength with the substrate [24, 30, 31]. Figure 3.6 shows cross-sectional images of Ti6Al4V deposited onto Ti6Al4V substrates using different nozzle traverse speeds [24]. It can be clearly seen that the porosity increases as the nozzle traverse speed increases. Moreover, at the lowest speed shown in Fig. 3.8a, delamination of the deposit from the substrate is observed, suggesting low adhesion strength caused by high residual stresses [17, 24]. Delamination at low nozzle traverse speeds and high powder feed rates was also reported in other studies, and frequently occurred in the authors' experiments [17]. Another potential effect of nozzle traverse speed is heating of the deposit and/or substrate by to the high-temperature impinging jet. Lower traverse gun speed results in higher deposit and substrate temperatures, which contributes to the deposition process [24, 32] but in turn causes thermal stresses at the interface [33].

Currently, there are no general rules on how to select the nozzle traverse speed. However, according to existing studies and the experience of the authors', the use of low nozzle traverse speeds often results in high localized residual stresses between the deposit and substrate, and low adhesion strength. These are undesirable for the application of CSAM for the repair of damaged components [17]. Therefore, it is not recommended to use a low nozzle traverse speed for CSAM.

3.5 Nozzle Scanning Step

The build-up of a CSAM deposit is achieved through nozzle scanning over the entire target surface line by line and layer by layer according to a pre-defined trajectory. In other words, the deposit is produced through the stacking of many single-track deposits. The most frequently used scanning strategy is an overlapping strategy where two neighbouring single-tracks overlap each other, as shown in Fig. 3.7. The

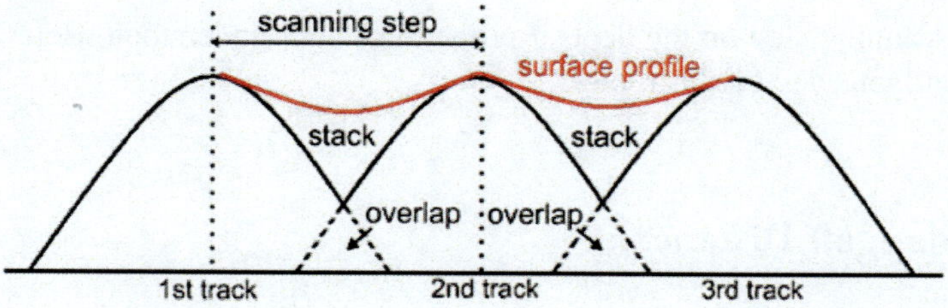

Fig. 3.7 Schematic of scanning step definition

scanning step is then defined as the interval between the centerlines of two single-track deposits. The main influence of scanning step on the deposit is the on the uniformity of the deposit thickness and the surface morphology; a properly chosen scanning step will lead to a homogeneous deposit and smooth surface.

The selection of scanning step depends upon the width of a single-track deposit. In the authors' experiments, the scanning step is always defined as half of the single-track width to guarantee the flatness of the deposit surface. However, a recent study has demonstrated that a more deliberate selection of scanning step can result in a less undulating surface. In that study, the optimum scanning step was estimated by using a self-developed software, Toolkit, [34]. Figure 3.8 shows the deposit thickness of cold sprayed aluminium as predicted by Toolkit and as measured experimentally, for two values of scanning step. Clearly, under the optimum scanning step (4 mm), the deposit thickness is more consistent than when using a larger scanning step (7 mm). This clearly shows the importance of scanning step in CSAM. However, investigations on the effects of varying scanning step are quite rare. The effect, if

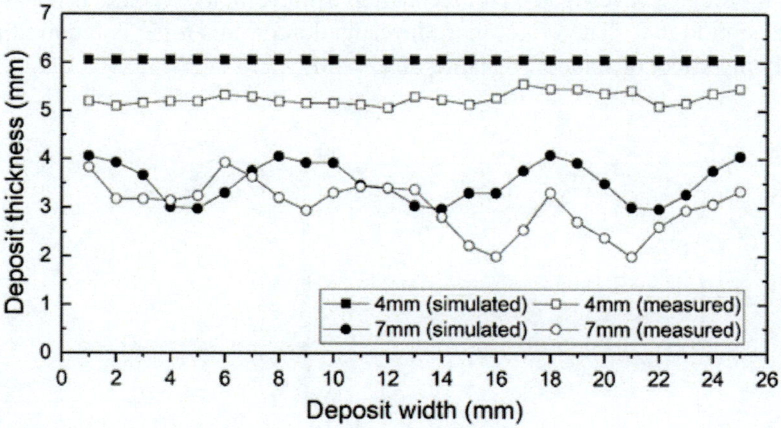

Fig. 3.8 Simulated surface flatness of aluminium deposit by Toolkit and measured one under the scanning steps of 4 and 7 mm. The width of the single-track is 16 mm. Data was picked from Ref. [34] and replotted

any, of scanning step on the deposit properties is not understood. Relevant studies are urgently needed on this topic.

3.6 Standoff Distance

Standoff distance is defined as the distance between nozzle outlet and the target surface. In a supersonic free jet, the intensity of the jet core decreases gradually along the jet central axis due to momentum exchange between the jet and the atmosphere [29, 35]. Inside the jet core, particles are subject to positive drag forces and thus rapidly accelerate. However, a short distance after the gas exits the diverging section of the nozzle, the gas decelerates to lower than the particle velocity and the resultant negative drag force causes the deceleration of particles. As a result, the particle impact velocity and deposition efficiency increase as the standoff distance increases from zero, but then reduces after an optimal operating point [36]. This phenomenon has been experimentally confirmed in cold sprayed aluminium, copper and titanium deposits in the work of Pattison et al. [36] and copper deposits in the work of Li et al., as shown in Fig. 3.9 [37]. However, Li et al. reported a much shorter optimal standoff distance, and also a monotonic reduction of deposition efficiency in aluminium and titanium deposits as the standoff distance increased from 10 to 110 mm. Besides these two investigations, there are also other studies focusing on the effect of standoff distance [38–41], but a consensus on the effect of nozzle standoff distance in CSAM cannot be drawn due to the small number of studies. Although most of the existing work used 10–40 mm as the standoff distance for manufacturing, the effect of standoff distance on the cold sprayed deposit properties is not well understood. Even for the most thoroughly investigated parameter, deposition efficiency, there is no consistent conclusion to date. Therefore, more investigation on this topic is required to fully clarify the effect of standoff distance on CSAM.

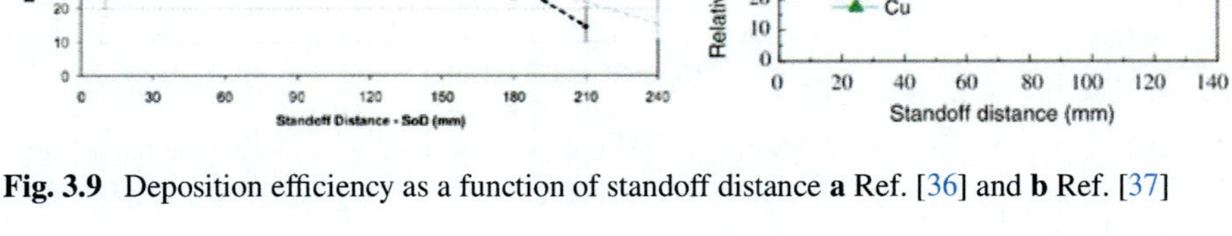

Fig. 3.9 Deposition efficiency as a function of standoff distance **a** Ref. [36] and **b** Ref. [37]

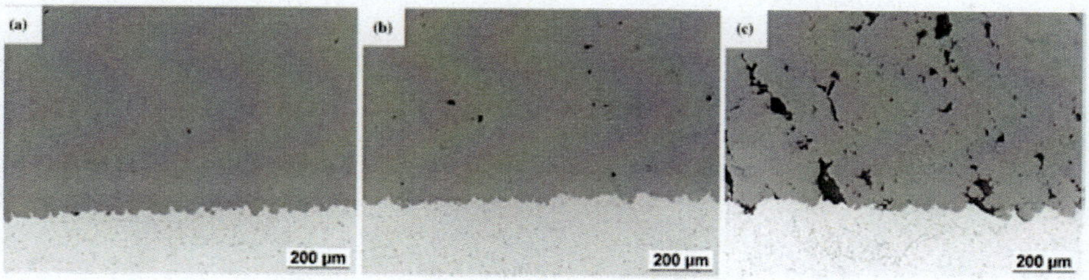

Fig. 3.10 Cross-sectional images of CSAM titanium deposit produced under the spray angle of **a** 90°, **b** 75° and **c** 45° [6]

3.7 Spray Angle

Spray angle is defined as the angle of the nozzle central axis to the target surface. It has a significant influence on the vector components of particle impact velocity, and consequently on the deposit properties. When spraying at an angle other than normal to the substrate surface, only the normal velocity component contributes to the particle deposition, while the tangential velocity component can play an opposing role by potentially detaching the deposited particles [42, 43]. As the spray angle decreases, the normal velocity component reduces but the tangential velocity component increases, which significantly deteriorates the deposit quality. Deposition efficiency, deposit strength and adhesion strength all reduce with decreasing spray angle [6, 38, 44–48]. This can be clearly noticed from Fig. 3.10 showing the increasing porosity in a titanium deposit with reducing spray angle [6]. In addition, modelling work has suggested that residual stresses decrease as the spray angle is reduced, but there is no experimental validation available so far [9].

Spray angle also affects the cross-sectional profile of single-track deposits. The profile of a track sprayed at an angle will have a skewed shape, as shown in Fig. 3.11. This is due to the inclined nozzle causing inconsistent standoff distance and thus non-uniform deposition efficiency over the particle impact area [25]. This may result in inhomogeneity of the deposit. Overall, angular spraying is harmful to the overall deposit performance; hence spraying at angles other than normal to the substrate is to be avoided in CSAM.

3.8 Nozzle Trajectory

Nozzle trajectory is defined as the path that nozzle takes during the manufacturing process. The basic principle of trajectory definition is to ensure homogeneous density and properties of the deposit. In general, when spraying on a flat or cylindrical surface, a simple round-trip (also known as a zig-zag pattern) trajectory is always applied. This trajectory is easy to define and able to yield a homogeneous deposit [30]. However, for a complex curved surface, trajectory definition should ensure that the

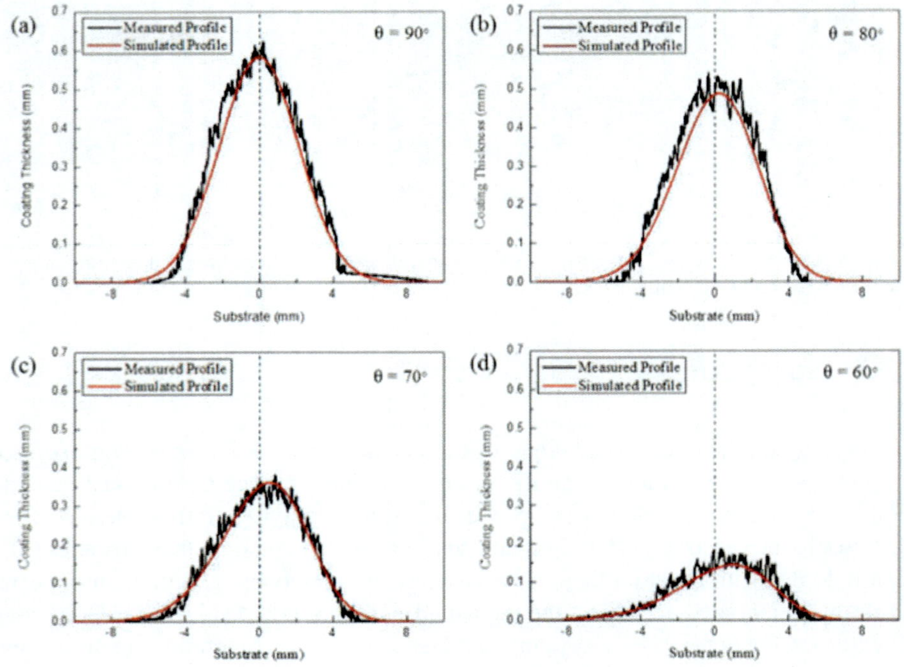

Fig. 3.11 Effect of spray angle on the single-track deposit cross-sectional profile [25]

standoff distance, spray angle, nozzle speed and other manufacturing parameters are kept constant over the curved surface to avoid any inhomogeneity of the deposited material. In addition, trajectory definition should also consider the curvature and transition between different features [49, 50]. Nozzle trajectories for complex surfaces are normally defined on an off-line programming platform by the aid of CAD/CAM software to accurately control the nozzle parameters and to adapt them to the surface topography [51]. Figure 3.12 shows a nozzle trajectory definition for depositing a turbine blade surface, and the resultant cold sprayed MCrAlY coating [52, 53]. The trajectory was first programmed on an off-line platform to define the nozzle traverse speed, scanning step, standoff distance and spray angle as shown in Fig. 12a. This trajectory was then executed by a robot arm, with the as-sprayed coating shown in Fig. 12b.

Another important structure that has been attracting attentions is thin-wall structure. As single-line cold sprayed deposits exhibit a Gaussian shape at the cross-section, it is not easy to fabricate such thin-wall structure directly through zig-zag trajectory [50, 54]. A novel building strategy developed in a recent study successfully fabricated vertical thin-wall structures through adjusting nozzle deflection angle (θ), offset distance (s) and retreat distance (d) as indicated in Fig. 3.13a [50]. Figure 3.13b shows the as-fabricated Cu thin-wall structures using the optimum building strategy. Nearly perfect thin-wall structures were successfully made through the optimum strategy.

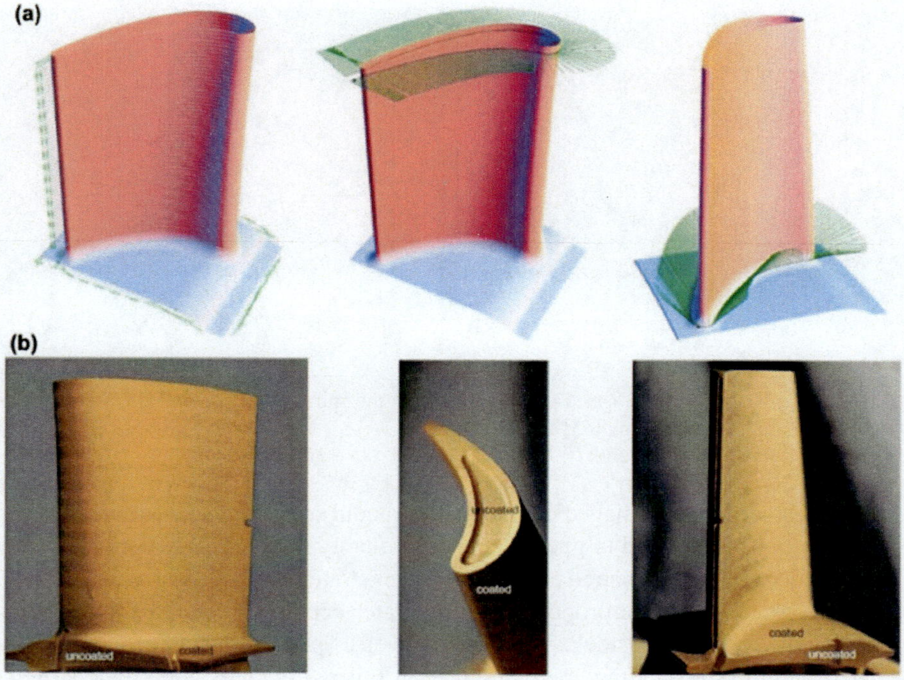

Fig. 3.12 **a** Nozzle trajectory for depositing a turbine blade surface and **b** the resultant cold sprayed MCrAlY coating [52, 53]

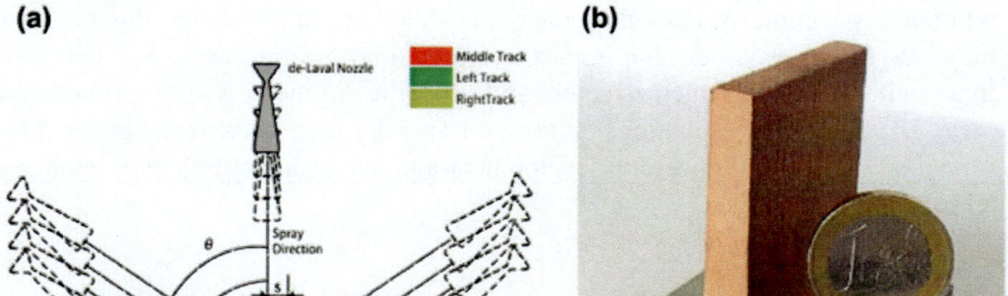

Fig. 3.13 **a** Schematic of the key parameters for nozzle trajectory design and **b** vertical thin-wall structures fabricated via CSAM [50]

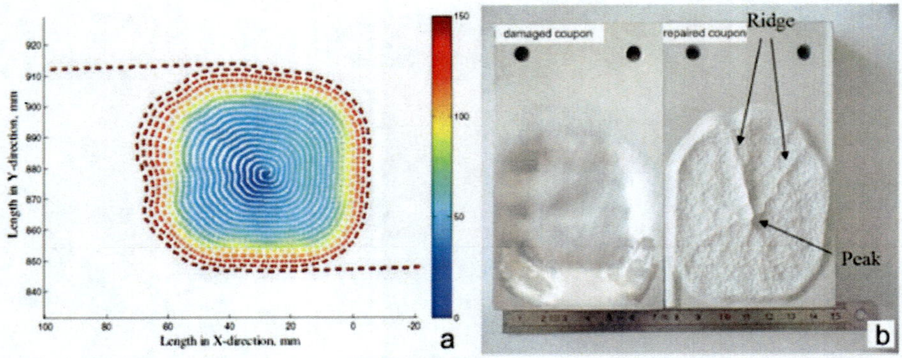

Fig. 3.14 **a** Nozzle traverse speed distribution along the spiral trajectory and **b** CSAM repaired coupon before final machining [55]

For restoration of damaged components via cold spray, the nozzle trajectory also needs to be well-defined to guarantee a high-quality repair. Figure 3.14 shows this spiral trajectory and a repaired coupon [55]. This strategy included varying the nozzle traverse speed in inverse proportion to the crater depth, to produce a homogenous deposit and to save feedstock. The nozzle traverse speed at the central area was set lower than the surrounding area, as shown in Fig. 3.14a. This will create a thicker deposit in the central deep area than in the surrounding shallow area, and thus the final repaired surface will be flat. The repaired coupon is shown in Fig. 3.14b, and clearly shows the advantages of this trajectory. The repaired surface is very flat without any significant redundant material in the deposit. However, this trajectory may also bring some negative impacts, as varied nozzle traverse speed will result in a nonhomogeneous microstructure and properties in the deposit as addressed in Sect. 3.4. To date investigations on nozzle trajectory have been very limited. More investigations on the trajectory development and use of novel trajectory strategies are strongly encouraged.

References

1. Yin, S., Meyer, M., Li, W., Liao, H., Lupoi, R.: Gas flow, particle acceleration, and heat transfer in cold spray: a review. J. Therm. Spray Technol. **25**, 1–23 (2016). https://doi.org/10.1007/s11666-016-0406-8
2. Huang, R., Fukanuma, H.: Study of the influence of particle velocity on adhesive strength of cold spray deposits. J. Therm. Spray Technol. **21**, 541–549 (2012). https://doi.org/10.1007/s11666-011-9707-0
3. Assadi, H., Kreye, H., Gärtner, F., Klassen, T.: Cold spraying—a materials perspective. Acta Mater. 116, 382–407 (2016). https://doi.org/10.1016/j.actamat.2016.06.034
4. Rokni, M.R., Nutt, S.R., Widener, C.A., Champagne, V.K., Hrabe, R.H.: Review of relationship between particle deformation, coating microstructure, and properties in high-pressure cold spray. J. Therm. Spray Technol. 1–48 (2017). https://doi.org/10.1007/s11666-017-0575-0

5. Goldbaum, D., Chromik, R.R., Yue, S., Irissou, E., Legoux, J.G.: Mechanical property mapping of cold sprayed Ti splats and coatings. J. Therm. Spray Technol. **20**, 486–496 (2011). https://doi.org/10.1007/s11666-010-9546-4
6. Binder, K., Gottschalk, J., Kollenda, M., Gärtner, F., Klassen, T.: Influence of impact angle and gas temperature on mechanical properties of titanium cold spray deposits. J. Therm. Spray Technol. **20**, 234–242 (2011). https://doi.org/10.1007/s11666-010-9557-1
7. Goldbaum, D., Shockley, J.M., Chromik, R.R., Rezaeian, A., Yue, S., Legoux, J.G., Irissou, E.: The effect of deposition conditions on adhesion strength of Ti and Ti6Al4V cold spray splats. J. Therm. Spray Technol. **21**, 288–303 (2012). https://doi.org/10.1007/s11666-011-9720-3
8. Luzin, V., Spencer, K., Zhang, M.X.: Residual stress and thermo-mechanical properties of cold spray metal coatings. Acta Mater. **59**, 1259–1270 (2011). https://doi.org/10.1016/j.actamat.2010.10.058
9. Shayegan, G., Mahmoudi, H., Ghelichi, R., Villafuerte, J., Wang, J., Guagliano, M., Jahed, H.: Residual stress induced by cold spray coating of magnesium AZ31B extrusion. Mater. Des. **60**, 72–84 (2014). https://doi.org/10.1016/j.matdes.2014.03.054
10. Cavaliere, P., Silvello, A.: Processing conditions affecting residual stresses and fatigue properties of cold spray deposits. Int. J. Adv. Manuf. Technol. **81**, 1857–1862 (2015). https://doi.org/10.1007/s00170-015-7365-y
11. Venkatesh, L., Chavan, N.M., Sundararajan, G.: The influence of powder particle velocity and microstructure on the properties of cold sprayed copper coatings. J. Therm. Spray Technol. **20**, 1009–1021 (2011). https://doi.org/10.1007/s11666-011-9614-4
12. Ghelichi, R., Bagherifard, S., Macdonald, D., Fernandez-Pariente, I., Jodoin, B., Guagliano, M.: Experimental and numerical study of residual stress evolution in cold spray coating. Appl. Surf. Sci. **288**, 26–33 (2014). https://doi.org/10.1016/j.apsusc.2013.09.074
13. Petráčková, K., Kondás, J., Guagliano, M.: Mechanical performance of cold-sprayed A357 aluminum alloy coatings for repair and additive manufacturing. J. Therm. Spray Technol. 1–10 (2017). https://doi.org/10.1007/s11666-017-0643-5
14. Eason, P.D., Kennett, S.C., Eden, T.J., Krull, I., Kowalski, B., Jones, J.L.: In situ observation of microstrain relief in cold-sprayed bulk copper during thermal annealing. Scr. Mater. **67**, 791–794 (2012). https://doi.org/10.1016/j.scriptamat.2012.07.029
15. Meyer, M.C., Yin, S., Mcdonnell, K.A., Stier, O., Lupoi, R.: Feed rate effect on particulate acceleration in cold spray under low stagnation pressure conditions. Surf. Coat. Technol. 237–245 (2016) (in review). https://doi.org/10.1016/j.surfcoat.2016.07.017
16. Meyer, M., Yin, S., Lupoi, R.: Particle in-flight velocity and dispersion measurements at increasing particle feed rates in cold spray. J. Therm. Spray Technol. **26**, 60–70 (2017). https://doi.org/10.1007/s11666-016-0496-3
17. Ozdemir, O.C., Widener, C.A., Carter, M.J., Johnson, K.W.: Predicting the effects of powder feeding rates on particle impact conditions and cold spray deposited coatings. J. Therm. Spray Technol. **26**, 1598–1615 (2017). https://doi.org/10.1007/s11666-017-0611-0
18. Taylor, K., Jodoin, B., Karov, J.: Particle loading effect in cold spray. J. Therm. Spray Technol. **15**, 273–279 (2006). https://doi.org/10.1361/105996306X108237
19. Pattison, J., Celotto, S., Morgan, R., Bray, M., O'Neill., W.: Cold gas dynamic manufacturing: A non-thermal approach to freeform fabrication. Int. J. Mach. Tools Manuf. 47, 627–634 (2007). https://doi.org/10.1016/j.ijmachtools.2006.05.001
20. Sova, A., Grigoriev, S., Okunkova, A., Smurov, I.: Potential of cold gas dynamic spray as additive manufacturing technology. Int. J. Adv. Manuf. Technol. **69**, 2269–2278 (2013). https://doi.org/10.1007/s00170-013-5166-8
21. Fang, D., Deng, S., Liao, H., Coddet, C.: The effect of robot kinematics on the coating thickness uniformity. J. Therm. Spray Technol. **19**, 796–804 (2010). https://doi.org/10.1007/s11666-010-9470-7
22. Deng, S., Liang, H., Cai, Z., Liao, H., Montavon, G.: Kinematic optimization of robot trajectories for thermal spray coating application. J. Therm. Spray Technol. **23**, 1382–1389 (2014). https://doi.org/10.1007/s11666-014-0137-7

23. Kotoban, D., Grigoriev, S., Okunkova, A., Sova, A.: Influence of a shape of single track on deposition efficiency of 316L stainless steel powder in cold spray. Surf. Coat. Technol. **309**, 2–9 (2016). https://doi.org/10.1016/j.surfcoat.2016.10.052
24. Tan, A.W.-Y.Y., Sun, W., Phang, Y.P., Dai, M., Marinescu, I., Dong, Z., Liu, E.: Effects of traverse scanning speed of spray nozzle on the microstructure and mechanical properties of cold-sprayed Ti6Al4V coatings. J. Therm. Spray Technol. **26**, 1484–1497 (2017). https://doi.org/10.1007/s11666-017-0619-5
25. Chen, C., Xie, Y., Verdy, C., Liao, H., Deng, S.: Modelling of coating thickness distribution and its application in offline programming software. Surf. Coat. Technol. (2016). https://doi.org/10.1016/j.surfcoat.2016.10.044
26. Seraj, R.A., Abdollah-zadeh, A., Dosta, S., Canales, H., Assadi, H., Cano, I.G.: The effect of traverse speed on deposition efficiency of cold sprayed Stellite 21. Surf. Coat. Technol. **366**, 24–34 (2019). https://doi.org/10.1016/j.surfcoat.2019.03.012
27. Champagne, V.K., Helfritch, D.J., Dinavahi, S.P.G., Leyman, P.F.: Theoretical and experimental particle velocity in cold spray. J. Therm. Spray Technol. **20**, 425–431 (2011). https://doi.org/10.1007/s11666-010-9530-z
28. Tabbara, H., Gu, S., McCartney, D.G., Price, T.S., Shipway, P.H.: Study on process optimization of cold gas spraying. J. Therm. Spray Technol. **20**, 608–620 (2011). https://doi.org/10.1007/s11666-010-9564-2
29. Yin, S., Liu, Q., Liao, H., Wang, X.: Effect of injection pressure on particle acceleration, dispersion and deposition in cold spray. Comput. Mater. Sci. **90**, 7–15 (2014). https://doi.org/10.1016/j.commatsci.2014.03.055
30. Rech, S., Trentin, A., Vezzù, S., Vedelago, E., Legoux, J.-G.G., Irissou, E.: Different cold spray deposition strategies: single- and multi-layers to repair aluminium alloy components. J. Therm. Spray Technol. **6061**, 1237–1250 (2014). https://doi.org/10.1007/s11666-014-0141-y
31. Moridi, A., Hassani Gangaraj, S.M., Vezzu, S., Guagliano, M.: Number of passes and thickness effect on mechanical characteristics of cold spray coating. Procedia Eng. **74**, 449–459 (2014). https://doi.org/10.1016/j.proeng.2014.06.296
32. Yin, S., Suo, X., Xie, Y., Li, W., Lupoi, R., Liao, H.: Effect of substrate temperature on interfacial bonding for cold spray of Ni onto Cu. J. Mater. Sci. **50**, 7448–7457 (2015). https://doi.org/10.1007/s10853-015-9304-6
33. Candel, A., Gadow, R.: Trajectory generation and coupled numerical simulation for thermal spraying applications on complex geometries. J. Therm. Spray Technol. **18**, 981–987 (2009). https://doi.org/10.1007/s11666-009-9338-x
34. Cai, Z., Deng, S., Liao, H., Zeng, C., Montavon, G.: The effect of spray distance and scanning step on the coating thickness uniformity in cold spray process. J. Therm. Spray Technol. **23**, 354–362 (2014). https://doi.org/10.1007/s11666-013-0002-0
35. Yin, S., Suo, X., Liao, H., Guo, Z., Wang, X.: Significant influence of carrier gas temperature during the cold spray process. Surf. Eng. **30**, 443–450 (2014). https://doi.org/10.1179/1743294414y.0000000276
36. Pattison, J., Celotto, S., Khan, a., O'Neill, W.: Standoff distance and bow shock phenomena in the cold spray process. Surf. Coat. Technol. 202, 1443–1454 (2008). https://doi.org/https://doi.org/10.1016/j.surfcoat.2007.06.065
37. Li, W.Y., Zhang, C., Guo, X.P., Zhang, G., Liao, H.L., Li, C.J., Coddet, C.: Effect of standoff distance on coating deposition characteristics in cold spraying. Mater. Des. **29**, 297–304 (2008). https://doi.org/10.1016/j.matdes.2007.02.005
38. Singh, R., Rauwald, K.H., Wessel, E., Mauer, G., Schruefer, S., Barth, A., Wilson, S., Vassen, R.: Effects of substrate roughness and spray-angle on deposition behavior of cold-sprayed Inconel 718. Surf. Coat. Technol. **319**, 249–259 (2017). https://doi.org/10.1016/j.surfcoat.2017.03.072
39. Won, J., Bae, G., Kang, K., Lee, C., Kim, S.J., Lee, K.A., Lee, S.: Bonding, reactivity, and mechanical properties of the kinetic-sprayed deposition of Al for a thermally activated reactive Cu liner. J. Therm. Spray Technol. **23**, 818–826 (2014). https://doi.org/10.1007/s11666-014-0088-z

40. Maledi, N.B., Oladijo, O.P., Botef, I., Ntsoane, T.P., Madiseng, A., Moloisane, L.: Influence of cold spray parameters on the microstructures and residual stress of Zn coatings sprayed on mild steel. Surf. Coat. Technol. **318**, 106–113 (2017). https://doi.org/10.1016/j.surfcoat.2017.03.062
41. Cavaliere, P., Silvello, A., Cinca, N., Canales, H., Dosta, S., Garcia Cano, I., Guilemany, J.M.: Microstructural and fatigue behavior of cold sprayed Ni-based superalloys coatings. Surf. Coat. Technol. **324**, 390–402 (2017). https://doi.org/10.1016/j.surfcoat.2017.06.006
42. Li, W.Y., Yin, S., Wang, X.F.: Numerical investigations of the effect of oblique impact on particle deformation in cold spraying by the SPH method. Appl. Surf. Sci. **256**, 3725–3734 (2010). https://doi.org/10.1016/j.apsusc.2010.01.014
43. Chen, C., Xie, Y., Yin, S., Planche, M.-P., Deng, S., Lupoi, R., Liao, H.: Evaluation of the interfacial bonding between particles and substrate in angular cold spray. Mater. Lett. **173**, 76–79 (2016). https://doi.org/10.1016/j.matlet.2016.03.036
44. Yin, S., Suo, X., Su, J., Guo, Z., Liao, H., Wang, X.: Effects of substrate hardness and spray angle on the deposition behavior of cold-sprayed Ti particles. J. Therm. Spray Technol. **23**, 76–83 (2013). https://doi.org/10.1007/s11666-013-0039-0
45. Gilmore, D.L., Dykhuizen, R.C., Neiser, R.A., Roemer, T.J., Smith, M.F.: Particle velocity and deposition efficiency in the cold spray process. J. Therm. Spray Technol. **8**, 576–582 (1999). https://doi.org/10.1361/105996399770350278
46. Li, C.-J., Li, W.-Y., Liao, H.: Examination of the critical velocity for deposition of particles in cold spraying. J. Therm. Spray Technol. **15**, 212–222 (2006). https://doi.org/10.1361/105996306X108093
47. Li, C.J., Li, W.Y., Wang, Y.Y., Yang, G.J., Fukanuma, H.: A theoretical model for prediction of deposition efficiency in cold spraying. Thin Solid Films **489**, 79–85 (2005). https://doi.org/10.1016/j.tsf.2005.05.002
48. Luo, X.T., Li, Y.J., Li, C.J.C.X., Yang, G.J., Li, C.J.C.X.: Effect of spray conditions on deposition behavior and microstructure of cold sprayed Ni coatings sprayed with a porous electrolytic Ni powder. Surf. Coat. Technol. **289**, 85–93 (2016). https://doi.org/10.1016/j.surfcoat.2016.01.058
49. Lynch, M.E., Gu, W., El-Wardany, T., Hsu, A., Viens, D., Nardi, A., Klecka, M.: Design and topology/shape structural optimisation for additively manufactured cold sprayed components. Virtual Phys. Prototyp. **8**, 213–231 (2013). https://doi.org/10.1080/17452759.2013.837629
50. Wu, H., Xie, X., Liu, M., Verdy, C., Zhang, Y., Liao, H., Deng, S.: Stable layer-building strategy to enhance cold-spray-based additive manufacturing. Addit. Manuf. **35**, 101356 (2020). https://doi.org/10.1016/j.addma.2020.101356
51. Yin, S., Aldwell, B., Lupoi, R.: Cold spray additive manufacture and component restoration, pp. 195–224 (2018)
52. Stier, O.: Economics of Cold Spray—Fundamental Cost Analysis. In: CSAM workshop, Worcester, USA (2012)
53. Stier, O.: Fundamental cost analysis of cold spray. J. Therm. Spray Technol. **23**, 131–139 (2013). https://doi.org/10.1007/s11666-013-9972-1
54. Vargas-Uscategui, A., King, P.C., Styles, M.J., Saleh, M., Luzin, V., Thorogood, K.: Residual stresses in cold spray additively manufactured hollow titanium cylinders. J. Therm. Spray Technol. **29**, 1508–1524 (2020). https://doi.org/10.1007/s11666-020-01028-3
55. Chen, C., Gojon, S., Xie, Y., Yin, S., Verdy, C., Ren, Z., Liao, H., Deng, S.: A novel spiral trajectory for damage component recovery with cold spray. Surf. Coat. Technol. (2016). https://doi.org/10.1016/j.surfcoat.2016.10.096

Chapter 4
Microstructures of Cold Sprayed Deposits

Abstract The microstructure of cold sprayed deposits directly determines the properties of CSAM products and thus it is of importance to well understand the microstructural evolution mechanism of cold sprayed deposits. The microstructure is largely dependent on the plastic deformation behaviour of cold spray feedstock particles during their deposition. As the plastic deformation of cold spray particles is not homogenous with severe plastic deformation occurring at the contact interfaces, cold sprayed deposits typically have heterogenous microstructure and grain structures. A large number of dislocations and newly formed grains are only localized in the shear zones at the interparticle interfaces, while the microstructure in the interior region is mainly characterized by coarse grains. This chapter thoroughly discusses the formation mechanism of dislocations and ultra fine grains (UFGs) in various cold sprayed deposits during cold spray deposition. The formation mechanisms of intermetallic compounds and amorphous phases are also interpreted and discussed.

4.1 Introduction

During cold spray process, powder particles impact onto substrate at an extremely high velocity. The kinetic energy of these particles transforms into thermal energy and strain energy upon impact, leading to the occurrence of severe plastic deformation (SPD) and adiabatic shear instability (ASI) at localised contact regions near the interface [1–4]. As a result, the oxide layer originally on the particle and substrate surfaces is broken up, which results in the intimate metal-to-metal contact and metallurgical bonding between particles and substrate. This has been widely regarded as the most prevailing hypothesis of the particle–substrate or interparticle bonding mechanism for cold spray [5, 6].

As compared with other SPD processes, such as equal channel angular pressing (ECAP) and friction stir processing (FSP), the particle deformation behaviour in cold spray process has its unique features. First, the deformation of the particles during cold spray deposition complete in an extremely short time scale (typically < 100 ns) [7], and thus the strain rate of the deformation can reach as high as 10^6 to 10^9 s^{-1} which is thousands of times higher than that is experienced in ECAP or

S. Yin and R. Lupoi, *Cold Spray Additive Manufacturing*, Springer Tracts in Additive Manufacturing, https://doi.org/10.1007/978-3-030-73367-4_4

FSP [6, 8–10]. Secondly, the heating and cooling rates experienced by cold spray particles are considerably high, reaching 10^9 to 10^{10} k s^{-1}[8, 11]. Such high heating and cooling rates lead to the formation of more crystal defects and substructures in cold sprayed deposits than in the materials produced via other SPD processes. Thirdly, the SPD of cold spray particles is heterogenous and localised in the shear zones at the periphery of impact interfaces as revealed by the simulation results of a single copper particle impacting onto a copper substrate as shown Fig. 4.1 [12]. These factors work together, resulting in a unique microstructure for cold sprayed deposits. For briefly illustrating this special microstructure, Fig. 4.2 shows the EBSD result of the cross-section of an Al 6061 particle in cold sprayed deposit [13]. Note that the cross-section was perpendicular to the particle impact direction. As can be seen, the grain structure of a single cold spray particle is heterogenous. The grains at the central region of the particle are much coarser than those in the surrounding area

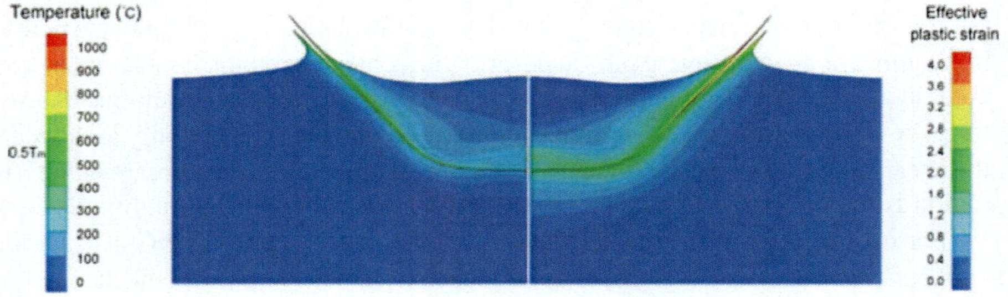

Fig. 4.1 Contours of the temperature and effective plastic strain of a single copper particle depositing on a copper substrate [12]

Fig. 4.2 EBSD images of the cross-section of an Al 6061 particle in cold sprayed deposit [13]

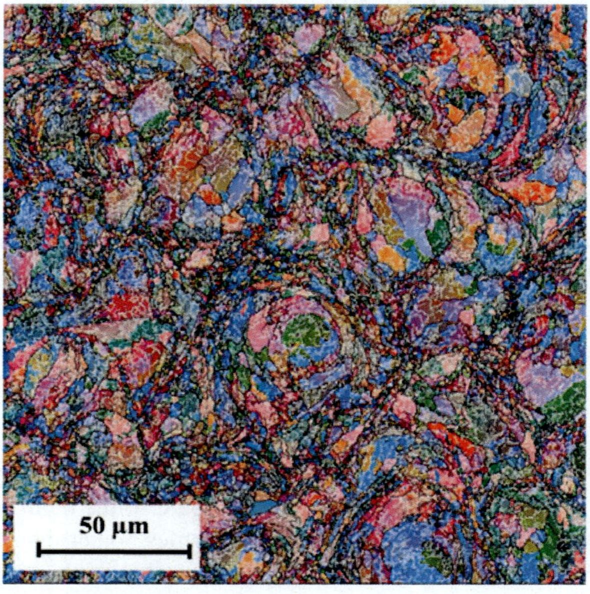

where elongated subgrains near the central zone and ultrafine grains (UFGs) marked by black dotted circle in the far surrounding region can be clearly observed. Apart from the heterogenous grain structures, the formation of intermetallic phases and amorphization at the localised interfacial regions further demonstrates the uniqueness and complexity of the microstructure of cold sprayed deposits.

4.2 Dislocations in Cold Sprayed Deposits

Among all kinds of crystal defects, dislocation is one of the most critical factors for inducing microstructure evolution [14]. Due to the ultra-high strain rate in cold spray, extensive dislocation multiplication occurs in deformed particles. The dislocation density in as-fabricated deposits can reach 10^{15} to 10^{16} m^{-2} [15–17], which is comparable to that in the highly deformed powders that have undergone high-energy ball milling [18]. Under such high dislocation density, a large number of dislocations accumulate and rearrange into different substructures, such as dislocation loops and cells. As the deformation progresses at localised interfacial regions, these substructures transform into subgrains which subsequently evolve into UFGs due to adiabatic temperature rise as schematised in Fig. 4.3 [19]. The increased dislocation density and the resultant new grain boundaries enhance the resistance of materials to deformation, and thus the hardness at highly deformed interfacial regions of a deposited particles is much higher than that of as-received powder particles. The behaviour of the dislocations in cold sprayed deposits have been extensively investigated through various methods such as X-Ray diffraction (XRD) [16, 20, 21], neutron diffraction and transmission electron microscope (TEM) [22]. Overall, the behaviour of the dislocations in cold sprayed deposits can be influenced by three factors: stack fault energy (SFE), alloying elements and manufacturing parameters.

4.2.1 Influence of Stack Fault Energy

Dislocations in high stack fault energy (SFE) metals are easier to migrate through climb and cross slip [23, 24]. In other words, they are more prone to be annihilated and rearranged into subgrain boundaries. On the contrary, dislocations in low SFE metals are difficult to slip and thus the formation of subgrain boundaries is prohibited. This leads to a low dislocation density in low SFE metals during deformation.

Table 4.1 lists the SFE of major cold spray materials that have been frequently used. For specifically interpreting the effect of SFE on the behaviour of the dislocations in cold sprayed deposits, aluminium and copper are selected as examples. Figure 4.4 shows the TEM images of cold sprayed Al 6061 deposit on Al 6061 substrate at the particle–substrate interface [25]. From Fig. 4.4a, an obvious microstructural variation can be observed from the particle–substrate interface towards the inner part of the particle. In the region close to the contact interface,

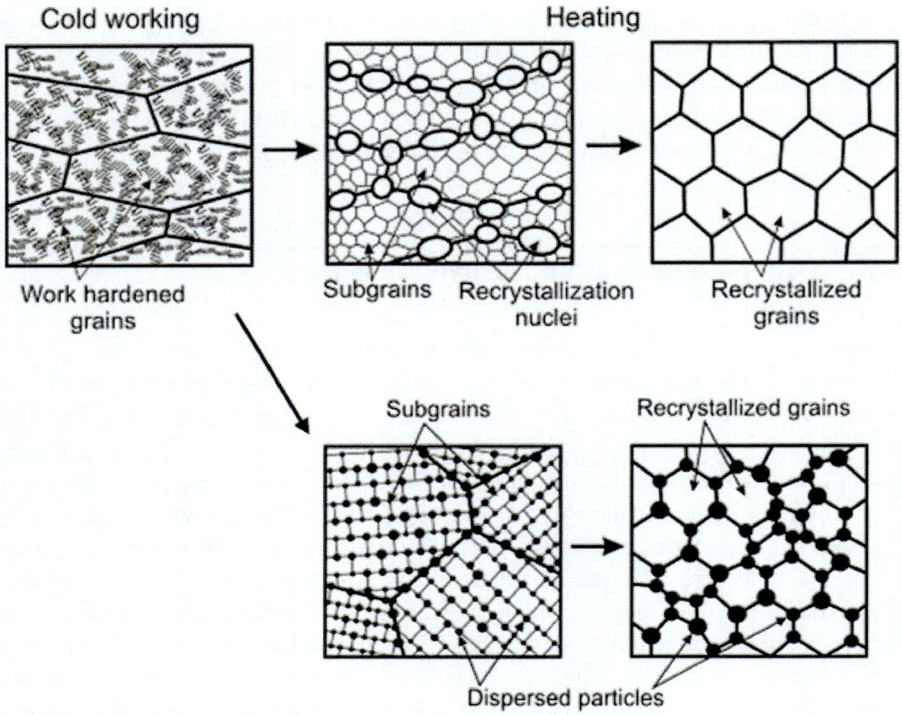

Fig. 4.3 Schematic representation of recrystallization procedure

Table 4.1 Stack fault energy of the metals typically used in cold spray [27–30]

Metal	γ_{SFE} (mJ/m^2)	Metal	γ_{SFE} (mJ/m^2)
Aluminium	166	Magnesium	125
Copper	78	Stainless steel 304	21
Nickel	128	Titanium	310

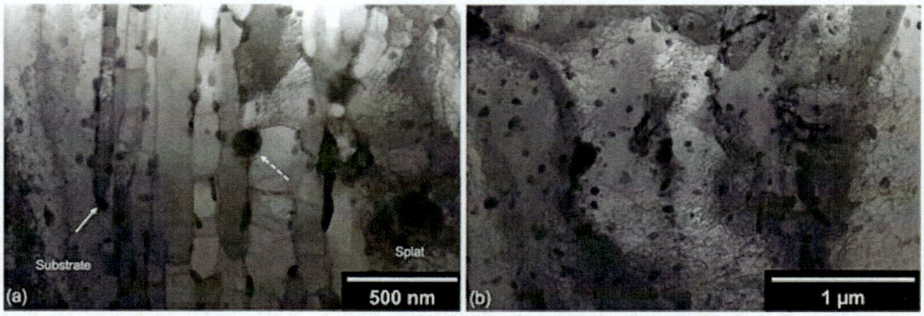

Fig. 4.4 TEM images showing the interfacial region between a deposited Al 6061 particle and Al 6061 substrate. **a** Recrystallized layer consisting of elongated grains in the substrate side and slightly coarsened ultrafine grains (UFGs) in the splat side, and **b** severely deformed cellular grains in a region in the splat next to the UFGs [25]

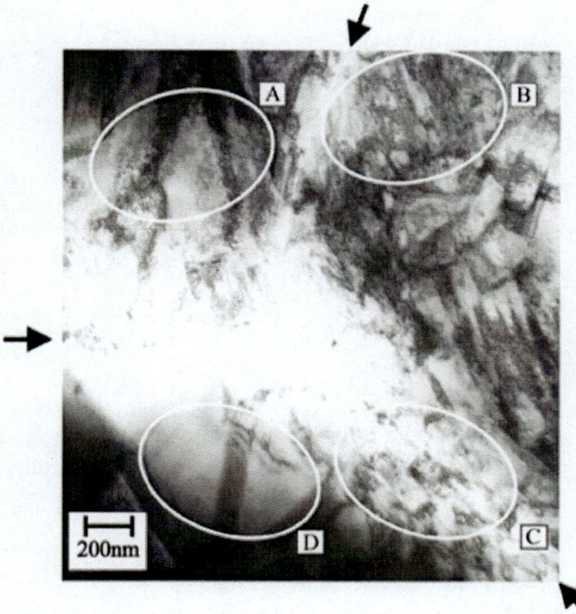

Fig. 4.5 TEM image showing the particle–particle interface in a cold sprayed copper deposit. Black arrows indicate the interfaces [26]

the elongated grains and UFGs having a size range between tens of to hundreds of nanometres can be observed. As the distance gets far away from the contact interface, grains become coarser and the number of dislocations also increases. The TEM image shown in Fig. 4.4b reveals that the dislocations are tangled to form continuous wall structures at the region next to the UFGs. The inner part of the wall structure is low of defect density, which indicates that subgrains have formed.

For comparison, Fig. 4.5 demonstrates the microstructure at the particle–particle interface of cold sprayed copper deposit (with lower SFE) [26]. Clearly, the grain structure of cold sprayed copper deposit is different from that of aluminium deposit. Overall, more dislocations can be found located near the contact interface as compared to aluminium deposit due to lower SFE. The dislocations are heavily tangled and form clear dislocation walls in the region A. UFGs can be found in the contact interface (region B and C). In addition, twinning structure which is an important deformation mode for FCC metals with low SFE can be found in region D.

4.2.2 Influence of Alloying Elements

Alloying elements in feedstock particles also play an important role in the behaivor of dislocations. It has been found that in cold sprayed Al-Cu alloy deposits, the disloction density in the deposit increases with increasing Cu content as shown in Fig. 4.6 [22]. This is because that the second phase particle or precipitation in the

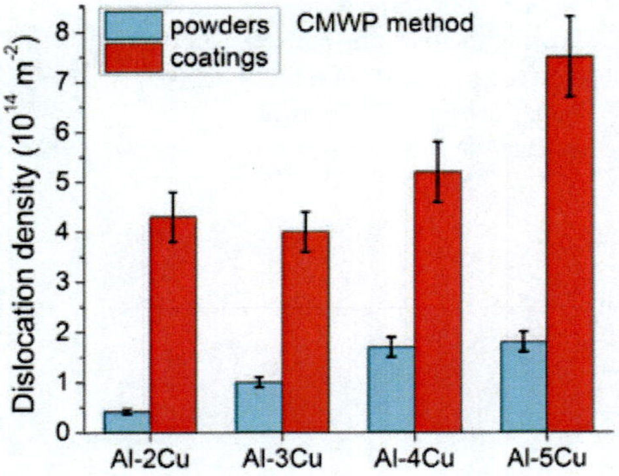

Fig. 4.6 Dislocation density of Al-Cu powders and deposits determined by the Convolutional multiple whole profile approach [22]

alloy hinders the migration of dislocations and grain bundaries. As a consequence, dislocation density is increased, and the size of substructures and grains are reduced [31, 32].

4.2.3 Influence of Manufacturing Parameters

Manufacturing parameters, particularly gas parameters, also affect the dislocation density and the formation of substructures. The parameters (e.g., higher gas temperature and pressure) that can lead to a higher particle impact vleocity bring in more strain and thermal energy to the deformed particles, which results in more severe plastic deformation and higher dislocation denisty. As an evidence, Fig. 4.7 shows the discolation density in cold sprayed tantalum deposits. It is seen that the dislocation density shows an increasing treand as the carrier gas temperature incresaes [33].

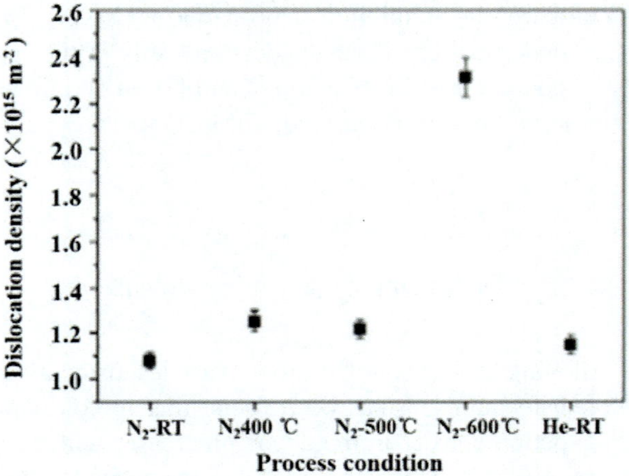

Fig. 4.7 Dislocation in cold sprayed tantalum deposits for different process condition [33]

4.3 Grain Structure

4.3.1 Grain Structure in a Single Splat Particle

In cold spray, single particle deposition trail is always used to analyse the microstructure evolution and bonding mechanism of cold sprayed deposits. In this section, copper, aluminium and nickel are selected as representative materials to describe the grain structures in cold sprayed deposits.

Figure 4.8 compares the grain structure of an as-received and an as-deposited cold sprayed copper powder [10]. The grain in the as-received particle is equiaxed and has a size range of 1–3 μm (Fig. 4.8a). After cold spray deposition, the grain structure exhibits considerable difference. The grain size gradually reduces from the particle top surface towards the contact interface, and such change is found related to the plastic deformation level. Based on the characteristics of the grain structure, the deposited particle can be divided into three regions. The grains in the Region A and B have a large size because the top and inner part of the particle do not experience too much plastic deformation. At the Region C and D, materials undergo extensive

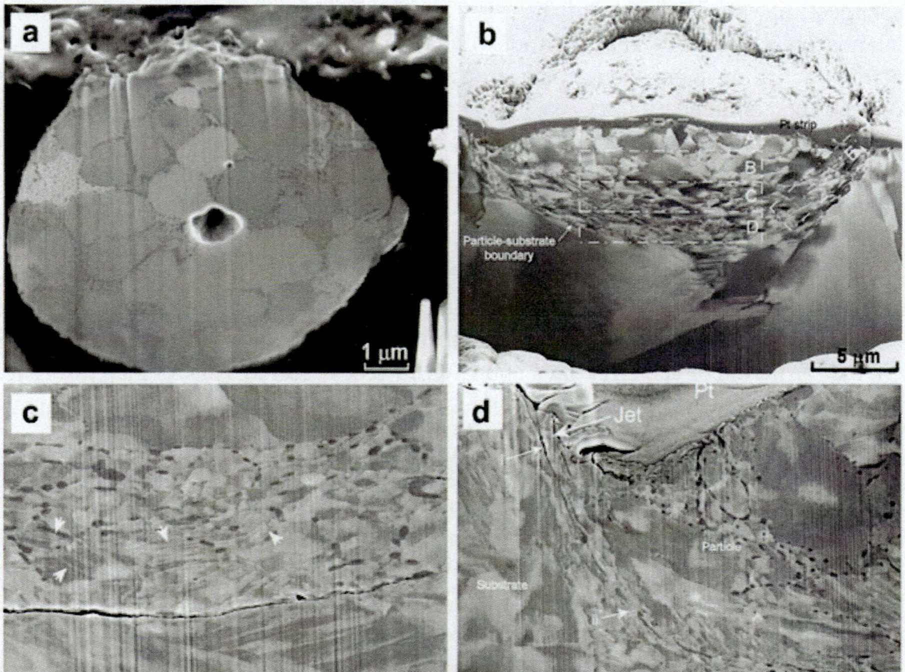

Fig. 4.8 SEM images of **a** an as-received copper particle and **b** a deposited copper particle onto copper substrate, and **c** magnified view of region D, and **d** magnified view of region E [10]

high-strain/strain-rate plastic deformation, and the grains are highly deformed and elongated to subgrains. Figure 4.8c, d show the SEM images revealing the grain structures at the Region D and E, respectively. It is clearly seen that equiaxed UFGs having a size range of 10–50 nm are formed near the contact interface. UFGs are typical grain structure for cold sprayed deposits. They are formed at the highly deformed regions near the impact interface through dynamic recrystallization, and is considered as one of the mechanisms resulting in the high hardness and strength of cold sprayed deposits. In addition, from Fig. 4.8d, twining structure (marked by white arrows) is also observed in the deposited particle. The formation of twining is an important deformation mechanism for low SFE metals and HCP metals, which can be enhanced under high temperature or low strain rate conditions. The grain structure distribution from the centre of the contact interface (Region D) towards the peripheral region (Region E) clearly reveals that the deformation at the periphery is higher than in the centre of the interface, which has also been confirmed by numerical modelling [8, 34–36]. More detailed TEM observation on particle–particle interface of copper deposits has already been shown in Fig. 4.5. Overall, the UFGs found in cold sprayed copper deposit are gathered in cluster and the size of the UFGs is closed to the scale of dislocation cell. Also, twining is formed in the strained grains, and there is no evidence of the formation of subgrains. All these features suggest that the formation of the UFGs grains in cold sprayed copper (low SFE) deposit is achieved by discontinuous dynamic recrystallization (DDRX) that has a clear nucleation stage and needs a strong accumulation of dislocations to enable the occurrence of nucleation.

As compared with copper, aluminium has a higher SFE energy, and thus the grain structure in a cold sprayed aluminium deposit is different from that in a copper deposit. The TEM images in Fig. 4.4 showing the interfacial region between a deposited Al 6061 particle and Al 6061 substrate have demonstrated that the grains near the contact interface are dramatically compacted into ribbon-like shape and is separated by HAGBs. At the inner region of the particle, the grain size is larger and the dislocations are tangled in high density. The accumulation of dislocations results in the formation of cell structures and LAGBs. These features are formed due to less deformation and lack of heat input into the inner region of the particle. Figure 4.9 shows the transmission EBSD results of a single Al-Cu particle depositing on steel 4340 substrate [17]. As shown in Fig. 4.9c, the ribbon-like grain structure in Fig. 4.4a are also found at the particle–substrate interface. It can be seen from the inverse pole figure (IPF) map in Fig. 4.9c that the misorientation between adjacent ribbon-like grains has no significant difference. Based on the grain structure and misorientation feature, it can be inferred that the ribbon-like nano-size grain is produced through geometrical dynamic recrystallisation (GDRX) mechanism. In GDRX, the DRX grains are formed through the following three steps: (i) the migration of HAGBs to form serrations, (ii) the reduction of grain thickness, and (iii) "pinching-off" of the serrated HAGBs when approaching 1–2 subgrain size distance. The size of GDRX grains is largely determined by the subgrain size, and the texture during GDRX process almost remains unchanged [37, 38].

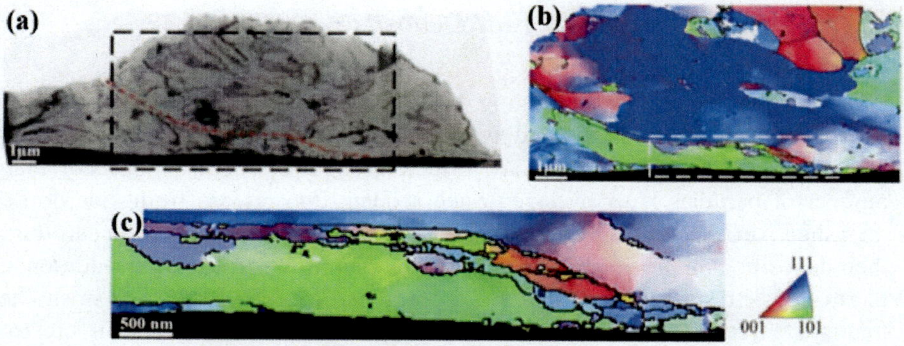

Fig. 4.9 Transmission EBSD results of a single Al-Cu particle depositing on steel 4340 substrate. **a** TEM image of the particle, **b** EBSD IPF map of the selected region from (**a**), and **c** magnified EBSD map of the region marked by white-dotted square in (**b**) [17]

Different from copper and aluminium, the grain structure in cold sprayed nickel deposit reveals an alternative DRX mechanism. Figure 4.10 shows the EBSD characterization of a cold sprayed nickel deposit. UFGs with size of tens of nanometres can be found next to coarser elongated grains. [39]. The misorientation of these UFGs is much higher than that of the coarser grains. This phenomenon indicates that the UFGs are formed through subgrains rotation mechanism of CDRX. For CDRX, there is no nucleation stage and the finer DRX is produced by (i) accumulation and rearrangement of dislocations to form subgrains and (ii) rotation of subgrains induced by continuous deformation [38]. The formation of DRX grains by subgrian rotation completes within extremely short time scale and the grains formed by this process are usually located at the HAGBs of initial grains.

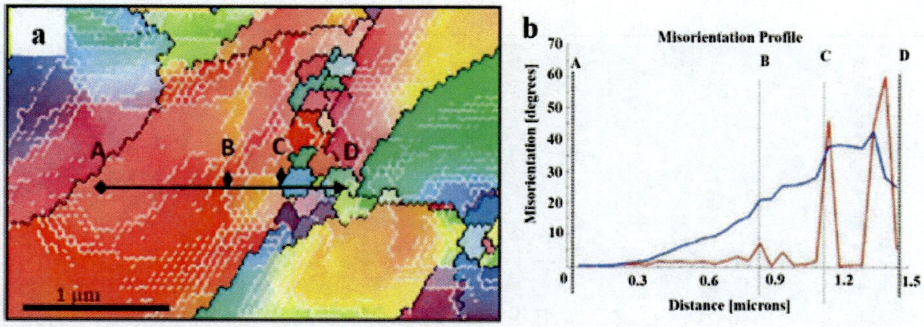

Fig. 4.10 **a** EBSD map of particle interface of cold sprayed nickel deposits, **b** Misorientation profile along the path ABCD. The blue curve in (**b**) indicates the point-to-point changes and the red curve indicates the point-to-origin changes [39]

4.3.2 Grain Structure in Full Deposits

Although single particle deposition trial is an effective method to study the microstructure evolution and bonding mechanism of cold sprayed deposits. In practical cold spray process, deposits are formed through continuous impacts of a large number of particles layer by layer. Such sequent impacts can further deform the deposited particles which experience little to no deformation in the top part during their deposition on previous layer, and change their microstructures. In addition, the thermal cycle resulted by continuous impact of particles also leads to an in-situ heat treatment effect on previously deposited layer, which also influences the microstructure evolution of deposited particles. Therefore, the grain structure evolution in cold sprayed full deposits is more complicated than in single deposited particle.

Figure 4.11 shows the cross-sectional EBSD misorientation map of a cold sprayed Al 6061 deposit [40]. The black dot in the map represents the crystalline area that is too small to be indexed and likely to be UFGs. It is seen that the grain structures at the interface between two particles are similar to those at the particle–substrate interface in single particle deposition trail. However, the grain structures in the joint of three particles are slightly complicated. A large un-indexed area can be found in the triple joint of three particles. Similar results can also be observed in cold sprayed copper deposit as revealed by the TEM observation in Fig. 4.5, where the triple joint of contact interface is filled with nanoscale equiaxed UFGs (Region C). This phenomenon may be the result of a more complicate deformation path induced by the impact of the third particle. Deformation path has been proven to have great influence on the microstructure in other SPD techniques [41].

During the continuous particle deposition process in cold spray, the top part of a deposited particle that does not experience deformation will be heavily deformed by the impact of the following particles and experience microstructure evolution. The

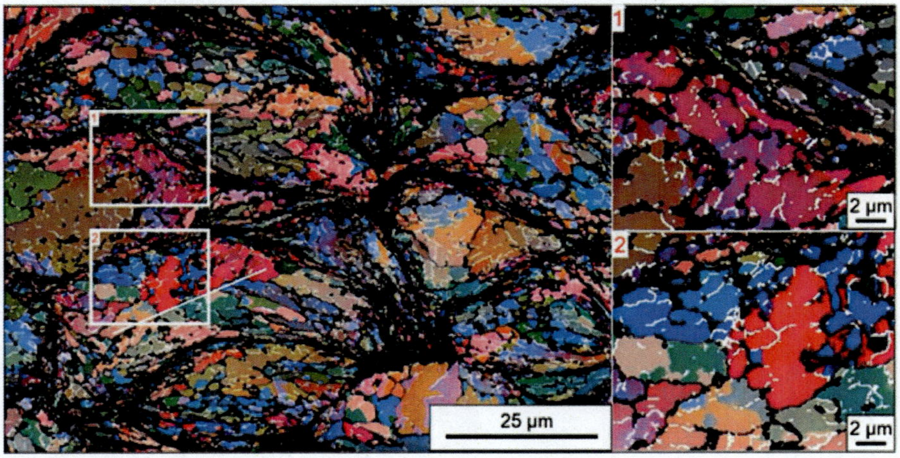

Fig. 4.11 Cross-sectional EBSD misorientation map of a cold sprayed Al 6061 deposit [40]

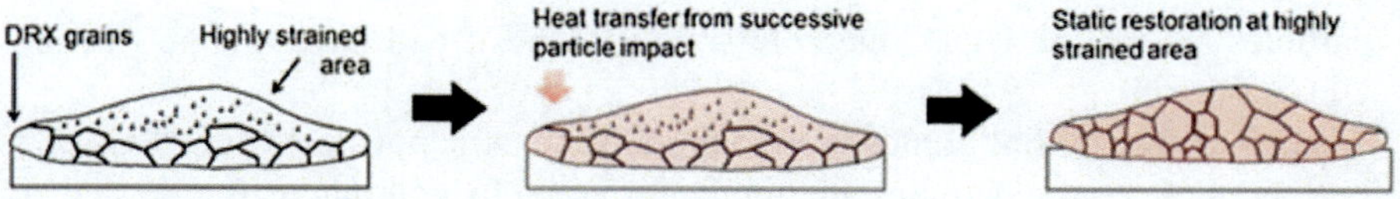

Fig. 4.12 Schematic of the microstructure evolution induced by SRV and SRX during the sequence particles impact in cold spray

resultant grain structures at these top regions are quite similar to the bottom regions. Apart from the top part, the strain in the inner region of the particle also increases owing to the tamping effect, which increases the strained energy stored in the particle. As schematised in Fig. 4.12, because of the thermal cycle induced by sequent impact and the annealing effect caused by post-deposition cooling, the stored energy is released through static recovery (SRV) and static recrystallisation (SRX), which can further refine the grain structure in cold sprayed deposit. SRV is a process that crystal defects in strained grains are annihilated or rearrange into different substructures. During SRX, refined RX grains are formed through nucleation and growth process, and the driven force is the difference of stored energy between two deformed grain. The SRX mechanism is similar to DDRX with the major difference being the absence of continuously strained energy input by deformation in SRX. Once the energy stored in deformed grains run out, SRX process will stop [42]. Typical refined grain structure induced by SRV and SRX in cold sprayed aluminium deposit can be found in Ref. [43], where the grains formed in this manner are equiaxed and free of defects. Typically, SRV usually takes place in the metals with high SFE and SRX is always found in low SFE metals.

4.4 Influential Factors for Grain Structure in Cold Sprayed Deposits

4.4.1 Microstructure in Feedstock Particles

UFGs in cold sprayed deposits are critical to enhance hardness and strength. Therefore, it is of importance to fabricate deposits with a large fraction of UFGs. Having small grains, the flow stress will be high, and this will elevate the stored energy in grains and provides more driving force for DRX. In addition, smaller grains normally indicate the existence of a large number of HAGBs. Thus, there will be more surface for the nucleation of DRX grains by DDRX and more subgrains at the HAGBs for rotation in CDRX. Up to now, various methods have been put into use to achieve smaller grain size in cold sprayed deposits. One of the feasible ways is to use powders with finer grain structure as feedstock. As most of the initial microstructure in as-received particles can be retained in final deposit, if the grains of starting powders particles are mostly UFGs, the cold sprayed deposit will have a large fraction of UFGs too.

High energy milling can be used to produce metallic powders with UFGs. In ball milling, the metallic particles are highly deformed by continuous impact and crash between metallic particles and milling balls (usually stainless steel or agate). Due to the SPD induced by high energy ball milling, nanocrystalline can be formed in the metallic powder particles. The average particles size of the powders also reduces quickly due to the crashing of particles. Rokin et al. [44] applied high energy milling to produce Al 5083 powder for cold spray. The as-received particles after the milling process have grains with a size of tens of nanometres as revealed by the TEM image and grain size statistics shown in Fig. 4.13. After deposition, the majority of the grains have an average size of 150 nm or less. In addition to the formation of nanocrystalline grain in starting powders, ball milling also increases the dislocation density in the particles through work hardening, and the high-density dislocations will be partially retained in the deposits. This results in a high driven force for dynamic or statistic restoration processes (including DRV, DRX, SRV and SRX) for deformed grains, which helps to further refine the grains during cold spray deposition [44–47]. It is also worth noting that although high-energy ball milling is able to produce fine grains in cold sprayed deposits, the accompanied work hardening effect significantly reduces the ductility of starting particles and deposition efficiency.

Apart from the initial microstructure in feedstock particles, adding alloying element into starting powders also influences the grain structure in cold sprayed deposit. During the deformation, second phase alloying elements could inhabit the motion of dislocation through Zener pining which can accelerate the local accumulation of dislocations. The rapid accumulation of dislocations will speed up the formation of subgrains and prompt LAGBs transferring into HAGBs. The presence of second phase also inhabits the migration of grain boundaries and slows down the growth of DRX grains that tend to grow abnormally owing to their low free energy after DRX process. Liu et al. studied the influence of alloying element through Al-Cu binary alloy system and found that the average grain size in the area of impact

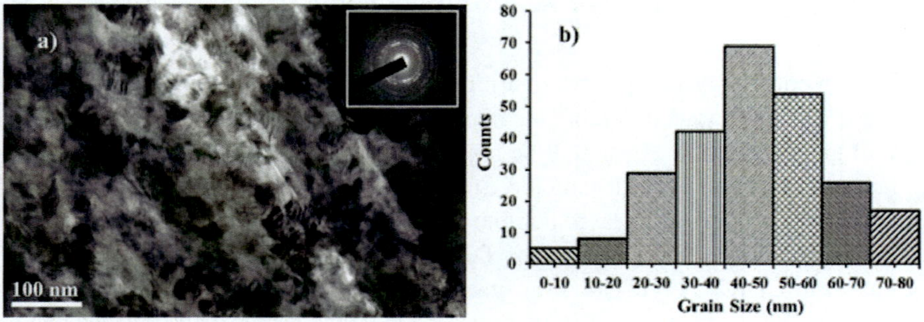

Fig. 4.13 **a** TEM image showing the grain structure in a ball-milled Al 5083, and **b** grain size distribution in the powder [44]

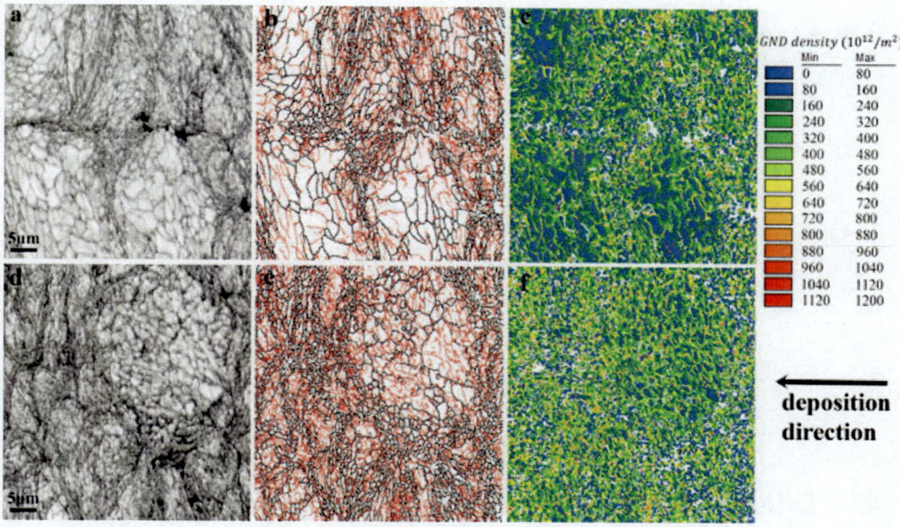

Fig. 4.14 EBSD results of the **a**–**c** cold sprayed Al-2Cu and **d**–**f** Al-5Cu deposits. (**a**, **d**) image quality maps, (**b**, **e**) grain boundary maps, HAGB (>10°) in black, LAGB (2° to 10°) in red, (**c**, **f**) geometrically necessary dislocation density maps [22]

interface had a clear decreasing tendency with the increase of Cu content as revealed by the EBSD analysis shown in Fig. 4.14 [22].

4.4.2 *Manufacturing Parameters*

Manufacturing parameters such as gas temperature, gas pressure and gas type also influence the extent of particle plastic deformation and thus grain structures in cold sprayed deposits. Table 4.2 summarises the deformation rate and the resultant grain size in different cold sprayed deposits. Typically, higher gas parameters cause higher particle impact velocity and larger particle plastic deformation which then promotes the formation of grain refinement. Apart from the high velocity, thermal energy also increases when gas temperature is increased, and this will enhance the SRV and SRX effect in the deposits [48–50]. In addition, particle and substrate preheating may also influence the grain structure evolution of cold sprayed deposits, but this will not be discussed in this book.

Table. 4.2 Summary of the deformation rate and grain size in different cold sprayed deposits

Metals	Initial grain size (μm)	Process parameter		Deformation rate	Final grain size (nm)	References
		Pressure (MPa)	Temperature (°C)			
Al 6061-T6	~	3	410	~	100–200	[25]
Al-Cu	2–3	1.21	225–275	0.57–0.6	<500	[17]
Cu	1.2	2.5	600		<100	[10]
Ni	1–5	3	600		100–200	[51]
Ta	1–2	3.2	800	–	<300	[52]
Ti	0.4–1	~	1000~	0.7	20~30	[53]
	~	3	600	0.25–0.3	<100	[54]

4.5 Intermetallic Compounds and Amorphization

4.5.1 Intermetallic Compounds

The clearance of oxide layer and the heating caused by plastic deformation during particle deposition can result in atomic diffusion and metallurgical bonding at the particle–particle or particle–substrate interfaces. In cold spray, materials have to be deposited onto substrates and in most cases the substrates are made of dissimilar materials with the feedstock powders [55]. In addition, cold spray has also been widely used to fabricate metal matrix composites (e.g., metal–metal, metal-ceramic and metal-polymer composites) [55]. In the fabrication of metal–metal composites, different kinds of metallic powder are mixed as the feedstock for cold spray deposition. These intimate contacts between different metals provide a chance for the formation of intermetallic compounds at the contact interfaces.

Normally, the deformation process of a cold spray particle completes within tens of nanoseconds. Therefore, although the thermal energy and crystalline defects caused by particle plastic deformation can prompt diffusion, the extremely short time period does not allow the formation large-area intermetallic compounds at the contact interfaces in cold spray. Figure 4.15 shows the TEM image of the interface between cold sprayed copper deposit and aluminium substrate and the EDX line scan across the interface [9]. It is seen that an $CuAl_2$ intermetallic layer with the thickness of up to 50 μm is formed at the interface. The formation of this intermetallic layer is likely due to the removal of oxide firms on the metal surface during deposition and the adiabatic temperature rise caused by the localised plastic deformation at the contact interface [9]. The intermetallic layer is also a clear evidence of the formation of metallurgical bonding. In some cases, local melting can occur in the deposit (e.g., aluminium, zinc and tin) as the result of high strain-rate deformation. In liquid state, the movement of atoms become much easier and a thicker intermetallic layer can be formed. Figure 4.16 shows the TEM images of Mg–Al interfaces where intermetallic

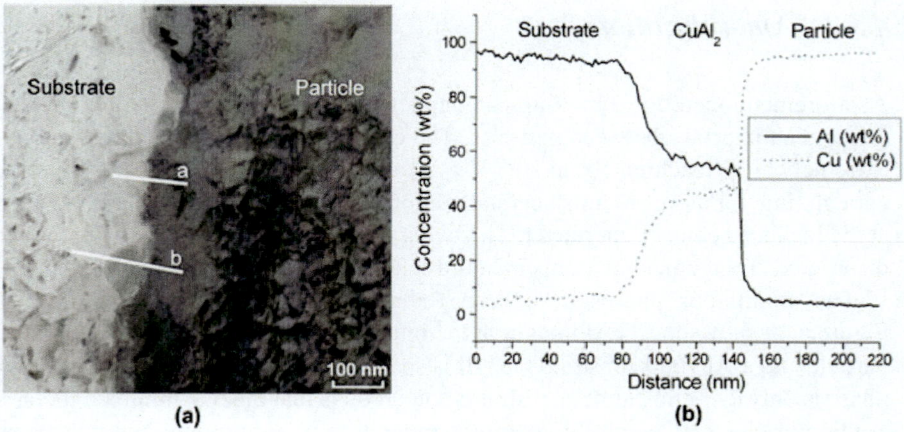

Fig. 4.15 **a** TEM image of the interface between cold sprayed copper deposit and aluminium substrate, and **b** the EDX line scan across the interface at region b [9]

Fig. 4.16 TEM image of the aluminium-magnesium interfaces in cold spray [56]

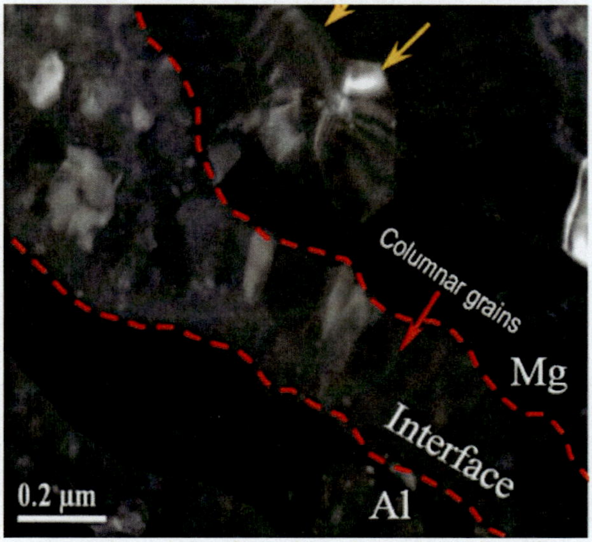

layer in the form of columnar grain structure (highlighted by red line) can be found [56]. It is also found that the thickness of intermetallic layer is affected by manufacturing parameters. For example, with the increase of gas pressure, the intermetallic layer likely increases due to the enhanced diffusion under higher temperature [57].

4.5.2 Amorphization

As aforementioned, local melting can happen at the highly deformed regions near the contact interface between particles. The cooling rate of these melting regions is extremely high, reaching 10^9 to 10^{10} K s^{-1}. Such rapid cooling rate does not provide enough time for atoms to finish crystallization and thus amorphous phase may form at the localised contact interface [57]. Owing to the extremely high cooling rate, the disordered arrangement of atoms in liquid state can be remained into solid state [58].

The amorphous phases have been found in various cold sprayed deposits. Figure 4.17 shows the TEM images at the interface between a deposited aluminium particle and a Al 1050 substrate [59]. It is seen that the thickness of the amorphous phase is only 2–5 nm. Numerical simulation predicts that the maximum temperature at the interface may reach the melting temperature of aluminium, and thus causes the formation of the amorphous phase. Actually, in most cases, local melting and columnar grains could be noticed in the regions where amorphous phase is found [59–62]. Therefore, it can be concluded that rapid quenching effect is one of the amorphization mechanism. Amorphous phases in metals are non-equilibrium and tend to crystallize if the temperature rises to a value over crystallization temperature. Different from certain alloy systems that are specially designed for amorphous phase, the amorphous phases are mainly found in cold sprayed metal and alloys that are difficult to form amorphous phase. Therefore, the amorphous phase in cold sprayed deposit is easy to recrystallize due to the thermal input brought by continuous deposition. Some researchers believe that the doping of other elements, for example the oxide element induced by broken oxide layer in deformation, can stabilize the amorphous in the interface [63].

Some studies also suggest that the rapid quenching may not be the only mechanism for amorphization. In the cold sprayed BN/NiCrAl composites, the amorphous layer was found at the interface between the BN and NiCrAl as shown in Fig. 4.18 [64]. However, the calculation of the peak temperature during the impact process indicates that the temperature was much lower than the melting point of NiCrAl alloy. In the case of cold sprayed aluminium onto magnesium substrate, amorphous layers were also detected in the aluminium-magnesium interface, and there was no evidence of any melting features [65]. The reasons for this kind of amorphization are provided below. In cold spray, the SPD at the impact interface results in a large number of crystal defects. When the density of defects is high enough, the atomic arrangement in solid-state will be disordered and similar to amorphous phase [66]. The following rapid cooling enables the retention of the transient microstructure with high crystal defects in the material after its cooling down. This amorphization mechanism is similar to that has been seen in the powders processed by high energy milling [67].

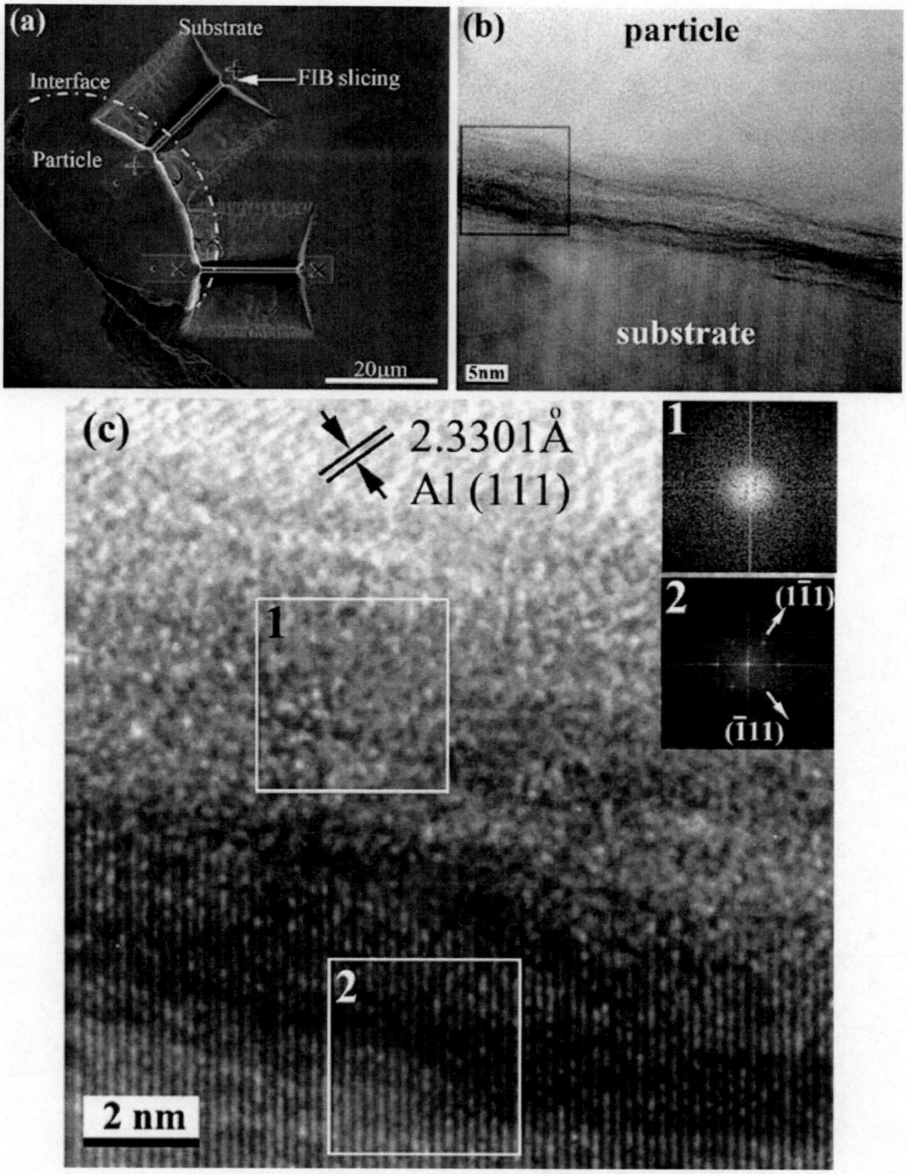

Fig. 4.17 TEM images at the interface between a deposited aluminium particle and an Al 1050 substrate. **a** preparation of the TEM sample, **b** high-resolution TEM image of particle–substrate interface, and **c** close observation at the selected box region and FFT patterns of selected area

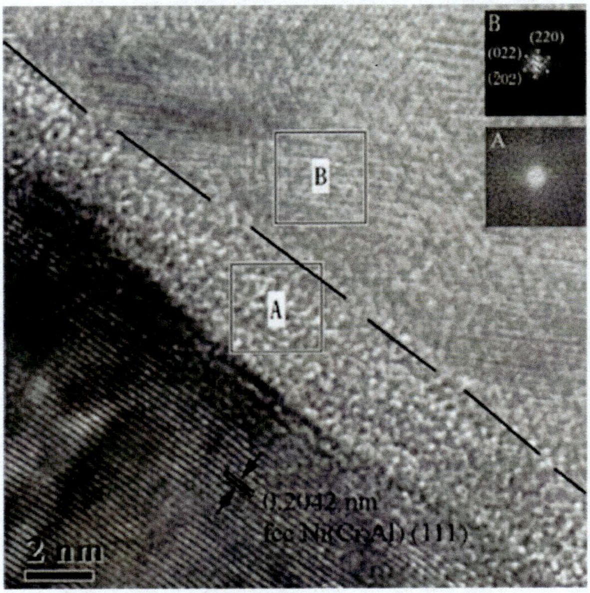

Fig. 4.18 High-resolution TEM images showing the formation of amorphous phase at the highly deformed interfaces of BN-NiCrAl [64]

References

1. Champagne, V.K. (ed.): Cold Spray Materials Deposition Process: Fundamentals and Applications. Woodhead Publ Ltd., Cambridge (2007)
2. Grujicic, M., Zhao, C.L.C., DeRosset, W.S.W., Helfritch, D.: Adiabatic shear instability based mechanism for particles/substrate bonding in the cold-gas dynamic-spray process. Mater. Des. **25**, 681–688 (2004). https://doi.org/10.1016/j.matdes.2004.03.008
3. Hines, J.A., Vecchio, K.S., Ahzi, S.: A model for microstructure evolution in adiabatic shear bands. Metall. Mater. Trans. A. **29**, 191–203 (1998)
4. Gouldstone, A., Choi, W.B., Chi, W., Wu, Y., Sampath, S.: The Cold Spray Materials Deposition Process: Fundamentals and Applications. In: Champagne, V.K. (ed.) (2007)
5. Assadi, H., Kreye, H., Gärtner, F., Klassen, T.: Cold spraying–a materials perspective. Acta Mater. **116**, 382–407 (2016)
6. Assadi, H., Gärtner, F., Stoltenhoff, T., Kreye, H.: Bonding mechanism in cold gas spraying. Acta Mater. **51**, 4379–4394 (2003). https://doi.org/10.1016/S1359-6454(03)00274-X
7. Hassani-Gangaraj, M., Veysset, D., Nelson, K.A., Schuh, C.A.: In-situ observations of single micro-particle impact bonding. Scr. Mater. **145**, 9–13 (2018). https://doi.org/10.1016/j.scriptamat.2017.09.042
8. Wang, Q., Ma, N., Takahashi, M., Luo, X., Li, C.: Development of a material model for predicting extreme deformation and grain refinement during cold spraying. Acta Mater. **199**, 326–339 (2020). https://doi.org/10.1016/j.actamat.2020.08.052
9. King, P.C., Bae, G., Zahiri, S.H., Jahedi, M., Lee, C.: An experimental and finite element study of cold spray copper impact onto two aluminum substrates. J. Therm. Spray Technol. **19**, 620–634 (2010). https://doi.org/10.1007/s11666-009-9454-7
10. King, P.C., Zahiri, S.H., Jahedi, M.: Microstructural refinement within a cold-sprayed copper particle. Metall. Mater. Trans. A Phys. Metall. Mater. Sci. **40**, 2115–2123 (2009). https://doi.org/10.1007/s11661-009-9882-5
11. Schmidt, T., Gärtner, F., Assadi, H., Kreye, H.: Development of a generalized parameter window for cold spray deposition. Acta Mater. **54**, 729–742 (2006). https://doi.org/10.1016/j.actamat.2005.10.005

12. Yin, S., Wang, X.F., Suo, X.K., Liao, H.L., Guo, Z.W., Li, W.Y., Coddet, C.: Deposition behavior of thermally softened copper particles in cold spraying. Acta Mater. **61**, 5105–5118 (2013). https://doi.org/10.1016/j.actamat.2013.04.041
13. Rokni, M.R., Widener, C.A., Crawford, G.A.: Microstructural evolution of 7075 Al gas atomized powder and high-pressure cold sprayed deposition. Surf. Coat. Technol. **251**, 254–263 (2014). https://doi.org/10.1016/j.surfcoat.2014.04.035
14. Rivera-Díaz-del-Castillo, E.I.G.-N., J.S., P.E.J.: Dislocation annihilation in plastic deformation: II. Kocks–Mecking analysis. Acta Mater. **60**, 2615–2624 (2012)
15. Grahle, P., Arzt, E.: Microstructural development in dispersion strengthened NiAl produced by mechanical alloying and secondary recrystallization. Acta Mater. **45**, 201–211 (1997). https://doi.org/10.1016/S1359-6454(96)00159-0
16. Rokni, M.R., Widener, C.A., Champagne, V.K., Crawford, G.A.: Microstructure and mechanical properties of cold sprayed 7075 deposition during non-isothermal annealing. Surf. Coat. Technol. **276**, 305–315 (2015). https://doi.org/10.1016/j.surfcoat.2015.07.016
17. Liu, T., Leazer, J.D., Brewer, L.N.: Particle deformation and microstructure evolution during cold spray of individual Al-Cu alloy powder particles. Acta Mater. **168**, 13–23 (2019). https://doi.org/10.1016/j.actamat.2019.01.054
18. Suryanarayana, C.: Mechanical alloying and milling. Prog. Mater. Sci. **46**, 1–184 (2001)
19. Sakai, T., Belyakov, A., Kaibyshev, R., Miura, H., Jonas, J.J.: Dynamic and post-dynamic recrystallization under hot, cold and severe plastic deformation conditions. Prog. Mater. Sci. **60**, 130–207 (2014). https://doi.org/10.1016/j.pmatsci.2013.09.002
20. Li, C.-J., Yang, G.-J., Gao, P.-H., Ma, J., Wang, Y.-Y., Li, C.-X.: Characterization of nanostructured WC-Co deposited by cold spraying. J. Therm. Spray Technol. **16**, 1011–1020 (2007). https://doi.org/10.1007/s11666-007-9096-6
21. Williamson, G.K., Hall, W.H.: X-ray line broadening from filed aluminium and wolfram. Acta Metall. **1**, 22–31 (1953). https://doi.org/10.1016/0001-6160(53)90006-6
22. Liu, T., Vaudin, M.D., Bunn, J.R., Ungár, T., Brewer, L.N.: Quantifying dislocation density in Al-Cu coatings produced by cold spray deposition. Acta Mater. **193**, 115–124 (2020). https://doi.org/10.1016/j.actamat.2020.04.040
23. Gourdet, S., Montheillet, F.: Effects of dynamic grain boundary migration during the hot compression of high stacking fault energy metals. Acta Mater. **50**, 2801–2812 (2002). https://doi.org/10.1016/S1359-6454(02)00098-8
24. Humphreys, J., Rohrer, G.S., Rollett, A.: Chapter 13—hot deformation and dynamic restoration. In: Humphreys, J., Rohrer, G.S., Rollett, A. (eds.) Recrystallization and Related Annealing Phenomena (Third Edition), pp. 469–508. Elsevier, Oxford (2017)
25. Evans, W.C., Dan, X., Houshmand, A., Müftü, S., Ando, T.: Microstructural characterization of aluminum 6061 splats cold spray deposited on aluminum 6061-T6 substrate. Metall. Mater. Trans. A Phys. Metall. Mater. Sci. **50**, 3937–3948 (2019). https://doi.org/https://doi.org/10.1007/s11661-019-05303-z
26. Borchers, C., Gärtner, R., Stoltenhoff, T., Kreye, H.: Microstructural bonding features of cold sprayed face centered cubic metals. J. Appl. Phys. **96**, 4288–4292 (2004). https://doi.org/10.1063/1.1789278
27. Gholinia, A., Humphreys, F.J., Prangnell, P.B.: Production of ultra-fine grain microstructures in Al–Mg alloys by coventional rolling. Acta Mater. **50**, 4461–4476 (2002)
28. Guo, Z., Miodownik, A.P., Saunders, N., Schillé, J.P.: Influence of stacking-fault energy on high temperature creep of alpha titanium alloys. Scr. Mater. **54**, 2175–2178 (2006). https://doi.org/10.1016/j.scriptamat.2006.02.036
29. Murr, L.E.: Interfacial Phenomena in Metals and Alloys (1975)
30. Schramm, R.E., Reed, R.P.: Stacking fault energies of seven commercial austenitic stainless steels. Metall. Trans. A. **6**, 1345 (1975)
31. Humphreys, F.J., Kalu, P.N.: Dislocation-particle interactions during high temperature deformation of two-phase aluminium alloys. Acta Metall. **35**, 2815–2829 (1987). https://doi.org/10.1016/0001-6160(87)90281-1

32. Nes, E., Ryum, N., Hunderi, O.: On the Zener drag. Acta Metall. **33**, 11–22 (1985). https://doi.org/10.1016/0001-6160(85)90214-7
33. Kim, J., Bae, G., Lee, C.: Characteristics of kinetic sprayed Ta in terms of the deposition behavior, microstructural evolution and mechanical properties: effect of strain-rate-dependent response of Ta at high temperature. Mater. Charact. **141**, 49–58 (2018). https://doi.org/10.1016/j.matchar.2018.04.023
34. Hassani, M., Veysset, D., Sun, Y.C., Nelson, K.A., Schuh, C.A.: Microparticle impact-bonding modes for mismatched metals: from co-deformation to splatting and penetration. Acta Mater. **199**, 480–494 (2020). https://doi.org/10.1016/j.actamat.2020.08.038
35. Suresh, S., Lee, S.W., Aindow, M., Brody, H.D., Champagne, V.K., Dongare, A.M.: Mesoscale modeling of jet initiation behavior and microstructural evolution during cold spray single particle impact. Acta Mater. **182**, 197–206 (2020). https://doi.org/10.1016/j.actamat.2019.10.039
36. Li, W.-Y., Liao, H., Li, C.-J., Li, G., Coddet, C., Wang, X.: On high velocity impact of micro-sized metallic particles in cold spraying. Appl. Surf. Sci. **253**, 2852–2862 (2006). https://doi.org/10.1016/j.apsusc.2006.05.126
37. Kassner, M.E., Barrabes, S.R.: New developments in geometric dynamic recrystallization. Mater. Ence Eng. A. **410**, 152–155 (2005)
38. Humphreys, J., Rohrer, G.S., Rollett, A.: Chapter 14—continuous recrystallization during and after large strain deformation. In: Humphreys, J., Rohrer, G.S., Rollett, A. (eds.) Recrystallization and Related Annealing Phenomena (Third Edition), pp. 509–526. Elsevier, Oxford (2017)
39. Zou, Y., Qin, W., Irissou, E., Legoux, J.G., Yue, S., Szpunar, J.A.: Dynamic recrystallization in the particle/particle interfacial region of cold-sprayed nickel coating: electron backscatter diffraction characterization. Scr. Mater. **61**, 899–902 (2009). https://doi.org/10.1016/j.scriptamat.2009.07.020
40. Sabard, A., Hussain, T.: Inter-particle bonding in cold spray deposition of a gas-atomised and a solution heat-treated Al 6061 powder. J. Mater. Sci. **54**, 12061–12078 (2019). https://doi.org/10.1007/s10853-019-03736-w
41. Ion, S.E., Humphreys, F.J., White, S.H.: Dynamic recrystallisation and the development of microstructure during the high temperature deformation of magnesium. Acta Metall. **30**, 1909–1919 (1982). https://doi.org/10.1016/0001-6160(82)90031-1
42. Humphreys, J., Rohrer, G.S., Rollett, A.: Chapter 6—recovery after deformation. In: Humphreys, J., Rohrer, G.S., Rollett, A. (eds.) Recrystallization and Related Annealing Phenomena (Third Edition), pp. 199–244. Elsevier, Oxford (2017)
43. Kang, K., Won, J., Bae, G., Ha, S., Lee, C.: Interfacial bonding and microstructural evolution of Al in kinetic spraying. J. Mater. Sci. **47**, 4649–4659 (2012). https://doi.org/10.1007/s10853-012-6332-3
44. Rokni, M.R., Widener, C.A., Nardi, A.T., Champagne, V.K.: Nano crystalline high energy milled 5083 Al powder deposited using cold spray. Appl. Surf. Sci. **305**, 797–804 (2014). https://doi.org/10.1016/j.apsusc.2014.04.010
45. Hall, A.C., Brewer, L.N., Roemer, T.J.: Preparation of aluminum coatings containing homogenous nanocrystalline microstructures using the cold spray process. J. Therm. Spray Technol. **17**, 352–359 (2008)
46. Ajdelsztajn, L., Jodoin, B., Kim, G.E., Schoenung, J.M.: Cold spray deposition of nanocrystalline aluminum alloys. Metall. Mater. Trans. A Phys. Metall. Mater. Sci. **36**, 657–666 (2005). https://doi.org/10.1007/s11661-005-0099-y
47. Tria, S., Elkedim, O., Hamzaoui, R., Guo, X., Bernard, F., Millot, N., Rapaud, O.: Deposition and characterization of cold sprayed nanocrystalline NiTi. Powder Technol. **210**, 181–188 (2011). https://doi.org/10.1016/j.powtec.2011.02.026
48. Meng, X., Zhang, J., Zhao, J., Liang, Y., Zhang, Y.: Influence of gas temperature on microstructure and properties of cold spray 304SS coating. J. Mater. Sci. Technol. **27**, 809–815 (2011). https://doi.org/10.1016/S1005-0302(11)60147-3

49. Yin, S., Suo, X., Liao, H., Guo, Z., Wang, X.: Significant influence of carrier gas temperature during the cold spray process. Surf. Eng. **30**, 443–450 (2014). https://doi.org/10.1179/1743294414y.0000000276
50. Tang, J., Zhao, Z., Li, N., Qiu, X., Shen, Y., Cui, X., Du, H., Wang, J., Xiong, T.: Influence of feedstock powder on microstructure and mechanical properties of Ta cold spray depositions. Surf. Coat. Technol. **377**, 124903 (2019). https://doi.org/10.1016/j.surfcoat.2019.124903
51. Moy, C.K.S., Cairney, J., Ranzi, G., Jahedi, M., Ringer, S.P.: Investigating the microstructure and composition of cold gas-dynamic spray (CGDS) Ti powder deposited on Al 6063 substrate. Surf. Coat. Technol. **204**, 3739–3749 (2010). https://doi.org/10.1016/j.surfcoat.2010.04.016
52. Koivuluoto, H., Honkanen, M., Vuoristo, P.: Cold-sprayed copper and tantalum coatings—detailed FESEM and TEM analysis. Surf. Coat. Technol. **204**, 2353–2361 (2010). https://doi.org/10.1016/j.surfcoat.2010.01.001
53. Kim, K., Watanabe, M., Kuroda, S.: Thermal softening effect on the deposition efficiency and microstructure of warm sprayed metallic powder. Scr. Mater. **60**, 710–713 (2009). https://doi.org/10.1016/j.scriptamat.2008.12.050
54. King, P.C., Busch, C., Kittel-Sherri, T., Jahedi, M., Gulizia, S.: Interface melding in cold spray titanium particle impact. Surf. Coat. Technol. **239**, 191–199 (2014). https://doi.org/10.1016/j.surfcoat.2013.11.039
55. Yin, S., Cavaliere, P., Aldwell, B., Jenkins, R., Liao, H.L., Li, W.Y., Lupoi, R.: Cold spray additive manufacturing and repair: fundamentals and applications. Addit. Manuf. **21**, 628–650 (2018). https://doi.org/10.1016/j.addma.2018.04.017
56. Shaha, S.K., Jahed, H.: Characterization of nanolayer intermetallics formed in cold sprayed Al powder on Mg substrate. Materials (Basel). **12**, 1317 (2019)
57. Lee, H., Shin, H., Ko, K.: Effects of gas pressure of cold spray on the formation of Al-based intermetallic compound. J. Therm. Spray Technol. **19**, 102–109 (2010). https://doi.org/10.1007/s11666-009-9407-1
58. Callister Jr, W.D., Rethwisch, D.G.: Fundamentals of Materials Science and Engineering: An Integrated Approach. Wiley (2020)
59. Xiong, Y., Kang, K., Bae, G., Yoon, S., Lee, C.: Dynamic amorphization and recrystallization of metals in kinetic spray process. Appl. Phys. Lett. **92**, 194101 (2008)
60. Li, C.-J., Li, W.-Y., Wang, Y.-Y.: Formation of metastable phases in cold-sprayed soft metallic deposit. Surf. Coat. Technol. **198**, 469–473 (2005). https://doi.org/10.1016/j.surfcoat.2004.10.063
61. Li, W.-Y., Li, C.-J., Yang, G.-J.: Effect of impact-induced melting on interface microstructure and bonding of cold-sprayed zinc coating. Appl. Surf. Sci. **257**, 1516–1523 (2010). https://doi.org/10.1016/j.apsusc.2010.08.089
62. Ko, K.H.H., Choi, J.O.O., Lee, H., Seo, Y.K.K., Jung, S.P.P., Yu, S.S.S.: Cold spray induced amorphization at the interface between Fe coatings and Al substrate. Mater. Lett. **149**, 40–42 (2015). https://doi.org/10.1016/j.matlet.2015.02.118
63. Xiong, Y., Xiong, X., Yoon, S., Bae, G., Lee, C.: Dependence of bonding mechanisms of cold sprayed coatings on strain-rate-induced non-equilibrium phase transformation. J. Therm. Spray Technol. **20**, 860–865 (2011). https://doi.org/10.1007/s11666-011-9634-0
64. Luo, X.-T., Yang, G.-J., Li, C.-J., Kondoh, K.: High strain rate induced localized amorphization in cubic BN/NiCrAl nanocomposite through high velocity impact. Scr. Mater. **65**, 581–584 (2011). https://doi.org/10.1016/j.scriptamat.2011.06.030
65. Wang, Q., Qiu, D., Xiong, Y., Birbilis, N., Zhang, M.X.: High resolution microstructure characterization of the interface between cold sprayed Al coating and Mg alloy substrate. Appl. Surf. Sci. **289**, 366–369 (2014). https://doi.org/10.1016/j.apsusc.2013.10.168
66. Branício, P.S., Rino, J.P.: Large deformation and amorphization of Ni nanowires under uniaxial strain: a molecular dynamics study. Phys. Rev. B—Condens. Matter Mater. Phys. **62**, 16950–16955 (2000). https://doi.org/10.1103/PhysRevB.62.16950
67. Weeber, A.W., Bakker, H.: Amorphization by ball milling. A review. Phys. B Condens. Matter. **153**, 93–135 (1988)

Chapter 5
Strengthening Strategies for Cold Sprayed Deposits

Abstract Cold sprayed deposits normally have unfavorable mechanical properties in their as-fabricated state, such as lower ductility, lower electrical conductivity, and lower thermal conductivity when compared to conventionally fabricated or fusion-based additive manufactured counterparts due to the inherent defects (i.e., porosity and unbonded interparticle boundary) found in the cold sprayed deposits. These defects generally exist in as-fabricated deposits, particularly when particle impact velocity is low. As structural weaknesses, defects have a negative impact on the mechanical properties of cold sprayed deposit. It is well known that increasing gas pressure and temperature can improve the mechanical properties of as-fabricated deposits. However, gas pressure and temperature cannot be infinitely increased due to the technological limitation of current cold spray systems. Fortunately, in recent years, some newly developed strengthening processes have been found effective to promote the mechanical properties of cold sprayed deposits. In this chapter, various strengthening strategies that have been frequently used for cold sprayed deposits, including conventional annealing, hot isostatic pressing, hot rolling, in-situ shot peening, in-situ densification, in-situ laser assistance, friction stirring and powder heat treatment will be introduced and discussed.

5.1 Conventional Annealing

Conventional annealing is one of the most effective and economical post-treatment processes to mitigate defects at the inter-particle boundaries by inducing recrystallization within deposited particles and across the inter-particle boundaries [1–12]. Figure 5.1 shows a comparison of the microstructure of an etched copper deposit in the as-fabricated and annealed states, respectively [13]. The inter-particle boundaries can be clearly seen in the as-fabricated sample shown in Fig. 5.1a. After annealing, recrystallization takes place both within the particles and across the inter-particle boundaries, helping to repair the defects at inter-particle boundaries. As a result of the annealing process, the grain structure evolves to fine equiaxed grains. The inter-particle boundaries become less visible and in some cases are completely eliminated, as shown in Fig. 5.1b [1, 14]. For copper, due to its excellent ductility and low

S. Yin and R. Lupoi, *Cold Spray Additive Manufacturing*, Springer Tracts in Additive Manufacturing, https://doi.org/10.1007/978-3-030-73367-4_5

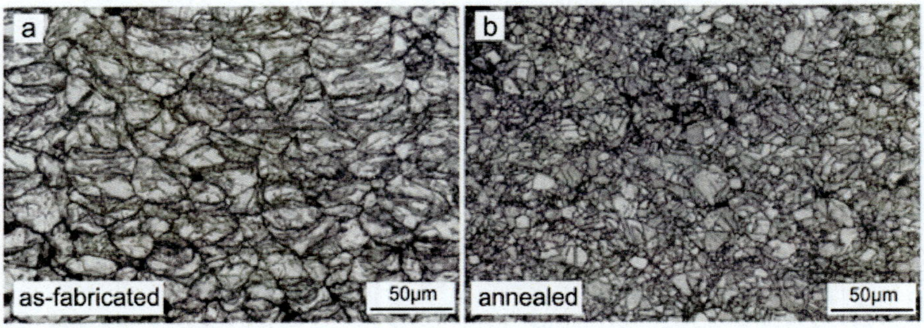

Fig. 5.1 Comparison of the microstructure of copper deposit **a** in the as-fabricated state and **b** in the annealed state. Annealing was performed at the temperature of 500 °C for 4 h [13]

strength to density ratio, the dominant defect that reduces the mechanical properties is the inter-particle boundaries. However, for other materials such as steel, titanium and its alloys, Inconel 718 and high entropy alloys, micro-pores are frequently found in the deposits due to their high hardness or high strength to density ratio. In these cases, both micro-pores and inter-particle boundaries affect the deposit properties. Figure 5.2 compares the microstructure between the as-fabricated and annealed stainless steel 316 deposits fabricated via CSAM [15]. The porosity reduced slightly from 3.6 to 2.6% after vacuum annealing treatment due to the occurrence of material diffusion across interparticle boundaries [15, 16]. Similar phenomenon can also been seen in materials such as titanium [17], titanium alloy [16], and Inconel [18].

For further explaining how annealing treatment improve the mechanical properties of cold sprayed deposits, copper as one of the most frequently used cold spray materials, is used as an example throughout. Table 5.1 summarizes the properties of cold sprayed copper deposits in their as-fabricated and after annealing states. For comparison, the properties of bulk copper are also provided in the table. As-fabricated deposits normally have inferior mechanical properties when compared with bulk materials, due to a large number of defects within the deposits. They are

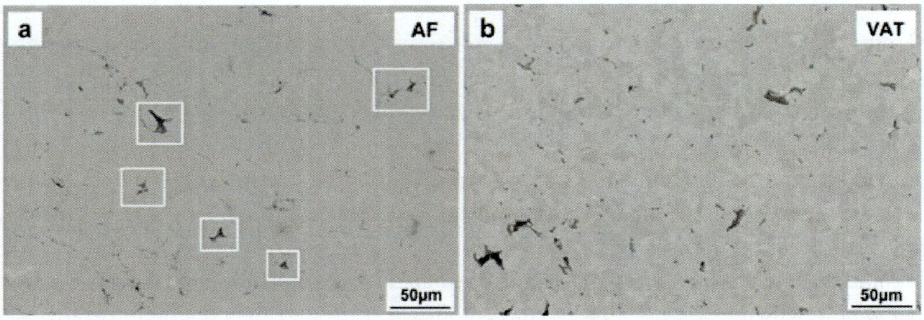

Fig. 5.2 Comparison of the microstructure of stainless steel 316 deposit **a** in the as-fabricated state and **b** in the annealed state. Annealing was performed at the temperature of 500 °C for 4 h [15]

Table 5.1 Summary of properties of CSAM copper deposits before and after heat treatment

Gas	Parameters		HT condition			Hardness, Hv		UTS, MPa		EL, %		EM, GPa		EC, MS/m		TC, W/m K		References
						50		225–385		45–5		118		58		394		
	T, °C	P, MPa	T, °C	t, h	PG	AF	HT	AF	HT	AF	HT	AF	HT	AF	HT	AF	HT	
N_2	800	3.0	300	4	Ar	–	–	294.6	338.4	0	0.68	84.3	104	–	–	–	–	[1]
			400	4	Ar		–		268.7		33.1		103		–		–	
			500	4	Ar		–		245.8		41.0		100		–		–	
			700	4	Ar		–		178.4		24.5		111		–		–	
N_2	500	3.0	200	5	Air	150.0	149.3	161.6	176.0	5.8	–	–	–	–	–	–	–	[2]
			300	5	Air		120.6		179.5		6.2		–		–		–	
			400	5	Air		94.1		200.0		7.4		–		–		–	
			500	5	Air		68.6		240.4		45.4		–		–		–	
N_2	305	3.0	200	1	Vacuum	152.9	97.7	46	86	< 0.1	0.12	–	–	34.9	40.8	–	–	[5, 6]
			400	1	Vacuum		77.9		171		0.51		–		49.0		–	
			600	1	Vacuum		73.6		210		8.03		–		51.4		–	
He	300	2.5	200	1	Vacuum	–	–	453	304	1.92	4.27	–	–	–	–	–	–	[5]
			400	1	Vacuum		–		278		20.99		–		–		–	
			600	1	Vacuum		–		262		23.85		–		–		–	
He	500	2.4	150	1	Air	108.0	90.3	320.0	295.8	2.4	4.0	–	–	–	–	–	–	[7]
			200	1	Air		85.1		256.8		15.8		–		–		–	
			300	1	Air		56.4		244.2		50.1		–		–		–	
N_2	300	3.0	200	1	–	150.1	98.1	–	–	–	–	–	–	35.7	41.5	–	–	[8]
			400	1	–		79.1		–		–				50.0		–	

(continued)

Table 5.1 (continued)

Gas	Parameters		HT condition			Hardness, Hv		UTS, MPa		EL, %		EM, GPa		EC, MS/m		TC, W/m K		References
						50		225–385		45–5		118		58		394		
	T, °C	P, MPa	T, °C	t, h	PG	AF	HT	AF	HT	AF	HT	AF	HT	AF	HT	AF	HT	
			600	1	–		74.8		–		–				52.8		–	
N_2	450	2.2	300	1	Air	122.6	–	–	–	–	–	–	–	17.6	29.7	–	–	[9]
			450	1	Air		–		–		–		–		32.3		–	
			300	4	Air		–		–		–		–		44.9		–	
			450	4	Air		–		–		–		–		45.1		–	
			300	1	Vacuum		86.7		–		–		–		49.3		–	
			450	1	Vacuum		–		–		–		–		49.6		–	
Air	400	2.0	400	1	Vacuum	169.3	109.1	–	–	–	–	82.7	94.0	–	30.2	–	–	[10, 11]
			600	1	Vacuum		93.81		–		–		99.4		33.4		–	
			800	1	Vacuum		72.4		–		–		115.0		36.5		–	
Air	400	0.6	200	1	Vacuum	99.5	75.3	–	–	–	–	–	–	–	–	125.2	164.5	
			400	1	Vacuum		74.0		–		–		–		–		262.6	[12]
			600	1	Vacuum		55.3		–		–		–		–		320.6	
N_2	400	3.0	400	2	Ar	145	80	–	–	–	–	–	–	38.6	52.7	–	–	[3]
Air	540	0.6	400	2	Ar	106	62	–	–	–	–	–	–	26.7	40.0	–	–	
Air	600	0.9	400	2	H_2	235.7	76.5	–	–	–	–	–	–	29.3	48.6	–	–	[4]

T—temperature; P—pressure; t—time, PG—protective gas, AF—as-fabricated, HT—heat treatment, UTS—ultimate tensile stress, EL—elongation, EM—elastic modulus, EC—electrical conductivity, TC—thermal conductivity

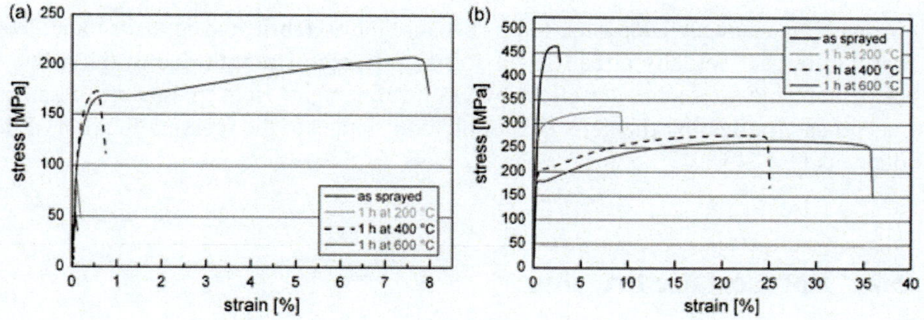

Fig. 5.3 Stress–strain curve of copper deposit before and after annealing: **a** produced by nitrogen, and **b** produced by helium [5]

very brittle and almost have no ductility. These effects are more prominent when the particle impact velocity is low [2, 5, 6]. Figure 5.3a shows the stress–strain curve of a copper deposit produced using nitrogen as a propulsive gas, in the as-fabricated and annealed states. The poor mechanical properties (yield strength, ultimate tensile strength and elongation) of the deposit in the as-fabricated state are clear [5]. In addition to the mechanical properties, other properties such as electrical conductivity [3, 4, 6, 8–11] and thermal conductivity [12] are also typically lower when compared to those of bulk materials. This is due to defects such as particle boundaries acting as barriers which prevent the transportation of current and heat through the deposits. Conversely, the hardness of the as-fabricated deposits is generally higher than that of bulk materials due to the work hardening effect which occurs during the deposition process [2–4, 6–12].

By annealing the as-fabricated deposits, both tensile strength and ductility are significantly improved, due to the mitigation of the defects in the deposits, as shown in Fig. 5.3a [2, 5, 6]. Higher annealing temperature and longer annealing time tend to cause to more significant improvement in properties. The tensile strength and elongation of annealed copper deposits can reach the same level as the bulk copper under proper annealing conditions. Similarly, the elastic modulus also improves after annealing due to the decrease in deposit defects [1, 10, 11]. However, because of the occurrence of recrystallization during the annealing process, work-hardened deposits are softened. As a result, the hardness of annealed deposits is normally lower than that of as-fabricated ones [2–4, 6–12]. Furthermore, as the inter-particle boundaries are repaired, the barriers for the transportation of heat and current are eliminated. Hence, annealed CSAM deposits have better electrical conductivity [3, 4, 6, 8–11] and thermal conductivity [12]. For copper, annealed deposits can achieve more than 90% of the electrical and thermal conductivity levels of bulk material.

However, under high particle impact velocity, heat treatment will improve the ductility but reduce the tensile strength of CSAM deposits, which is quite similar to the heat treatment of bulk materials [1, 5, 7]. Figure 5.3b shows the stress–strain curve of a copper deposit in its as-fabricated state and annealed state produced with helium [5]. It is clearly seen that the as-fabricated copper deposit had higher tensile strength than annealed deposits. This is because the as-fabricated deposits under high particle impact velocity already gain sufficiently high strength due to the absence of defects and strengthening effect by grain refinement. Heat treatment, therefore, has no use for healing the defects but only contributes to the recrystallization of the deposits [19, 20].

5.2 Hot Isostatic Pressing

Hot isostatic pressing (HIP) is a process designed to increase the density of materials by reducing or eliminating their porosity. In this process, high pressure (up to 200 MPa) and high temperature (up to 2000 °C) are applied to the processed materials simultaneously in a pressurized vessel filled with inert gas. Under such extreme conditions, the materials are greatly compacted, which results in a significant improvement of density, interparticle bonding, and mechanical properties of the processed materials. As a novel thermo-mechanical processing technology, HIP has been successfully used as a heat treatment process for densifying and strengthening cold sprayed deposits.

Figure 5.4 shows the X-ray CT reconstructions of the pores in cold sprayed Ti6Al4V [21] and 316L stainless steel deposits [15]. HIP treatment significantly reduced the porosity of Ti6Al4V deposits from 2.44 to 1.50% (produced with nitrogen), and from 1.16 to 0.04% (produced with helium), and that of 316L deposit from 3.36 to 0.27%. These results clearly demonstrate that HIP treatment can effectively compact and densify cold sprayed deposits. In comparison with conventional air/vacuum annealing treatments, HIP performs much better in deposit densification

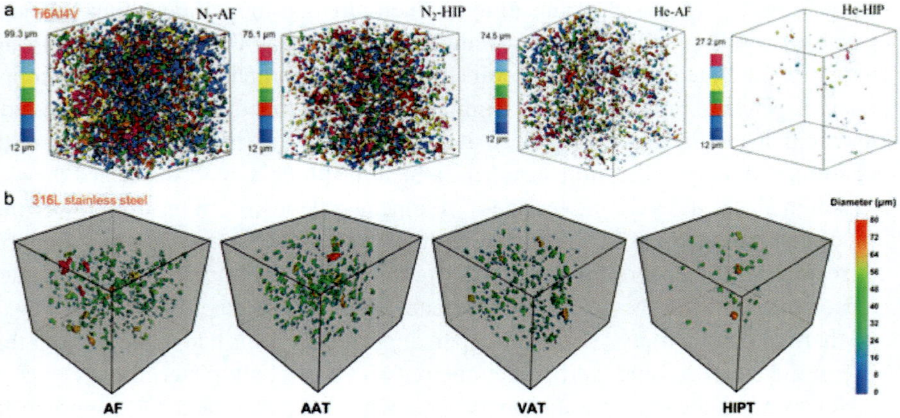

Fig. 5.4 X-ray CT reconstruction of the pores in cold sprayed **a** Ti6Al4V and **b** 316L stainless steel deposits [15, 21]. AF refers to as-fabricated; HIP refers to hot isostatic pressing; AAT refers to air annealing treatment; VAT refers to vacuum annealing treatment; HIPT refers to hot isostatic pressing treatment

as can be found in Fig. 5.4b. Similar results were also found in cold sprayed titanium deposit [22] and other Ti6Al4V deposit [23]. Apart from the densification effect, due to its high processing temperature, HIP processing also alters the phase and grain structures of cold sprayed deposits and heals the interparticle boundaries through high-temperature atomic diffusion, which is similar to what conventional annealing plays [15, 21, 23]. The SEM images shown in Fig. 5.5 demonstrate that the phase structure of as-fabricated Ti6Al4V deposits was changed from elongated α phase with fine-scale β phase to uniformly distributed β phase around by equiaxed α grain after HIP [21, 23].

As for the mechanical properties, HIP has been proved to significantly improve the strength and ductility of cold sprayed deposits. Figure 5.6 summarizes the ultimate strength and elongation at break of selected cold sprayed deposits before and after HIP. Clearly, for all deposits, their mechanical properties are greatly enhanced after HIP. An example showing the stress–strain curves of cold sprayed Ti6Al4V deposits before and after HIP is provided in Fig. 5.7. Although HIP is able to improve the

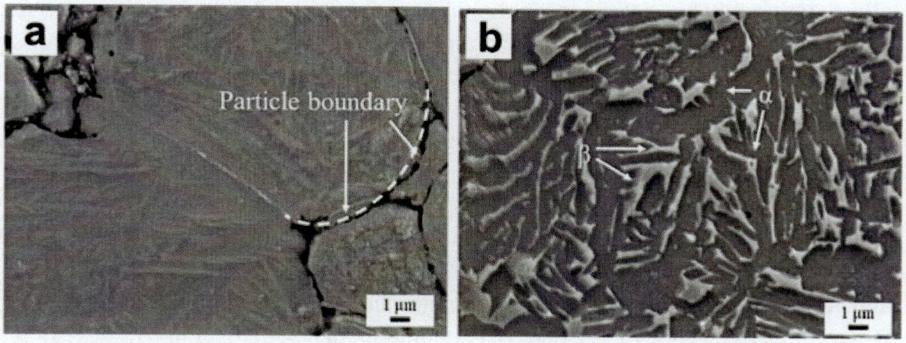

Fig. 5.5 Microstructures of cold spayed Ti6Al4V deposits **a** before and **b** after HIP treatment [21]

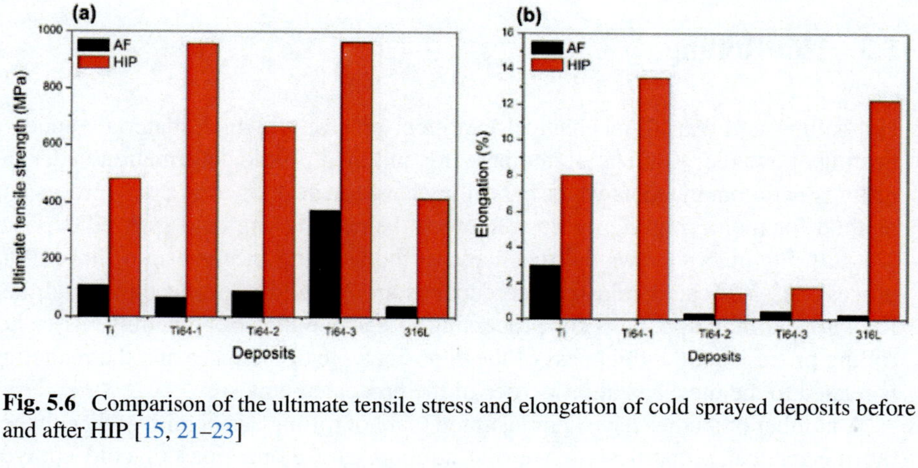

Fig. 5.6 Comparison of the ultimate tensile stress and elongation of cold sprayed deposits before and after HIP [15, 21–23]

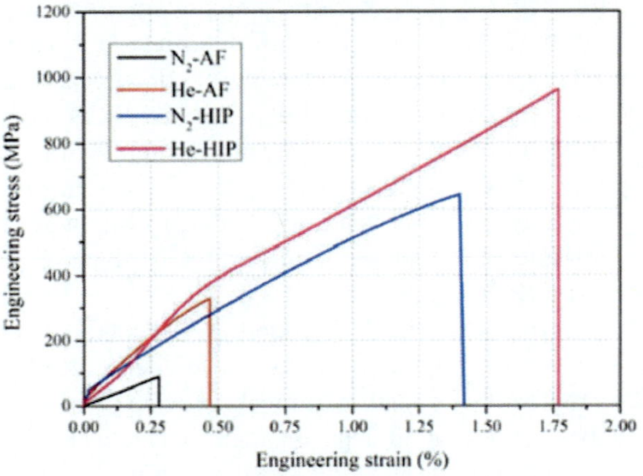

Fig. 5.7 Stress–strain curves of cold sprayed Ti6Al4V deposits before and after HIP [21]

mechanical properties of cold sprayed deposits, its strengthening mechanism is actually quite similar to that of conventional annealing. The improvement of mechanical properties is mainly attributed to the improved interparticle interfaces through enhanced atomic diffusion and grain structure change; the reduced porosity only slightly contributes to the strengthening. For this reason, HIP performs almost the same as annealing in terms of improving the strength and ductility of cold sprayed deposits [15].

Despite HIP is a promising technology for densifying and strengthening cold sprayed deposits, its high cost may limit its wide application for general cold spray products. In addition, it shows few to no advantages as compared to conventional annealing process in terms of improving strength and ductility as addressed above. However, it is worth noting that HIP treatment may greatly improve the fatigue strength of cold sprayed deposits due to the reduction of porosity.

5.3 Hot Rolling

Hot rolling is a thermo-mechanical treatment process in which materials undergo heating above recrystallization temperature and then plastic deformation under the rolling operation. In recent years, hot rolling has been developed as a post-processing method for modifying the microstructure and strengthening cold sprayed deposits [24–30]. Figure 5.8 shows the schematic of the working mechanism of hot rolling process. As-fabricated cold sprayed deposits are heated in a furnace first and then fed into a rolling apparatus for processing to achieve the strengthening. After hot rolling processing, the thickness of the deposits typically reduces and the reduction is related to the mechanical properties of the processed deposits.

A number of studies have demonstrated that hot rolling can significantly enhance the mechanical properties (i.e. tensile strength and elongation) of cold sprayed

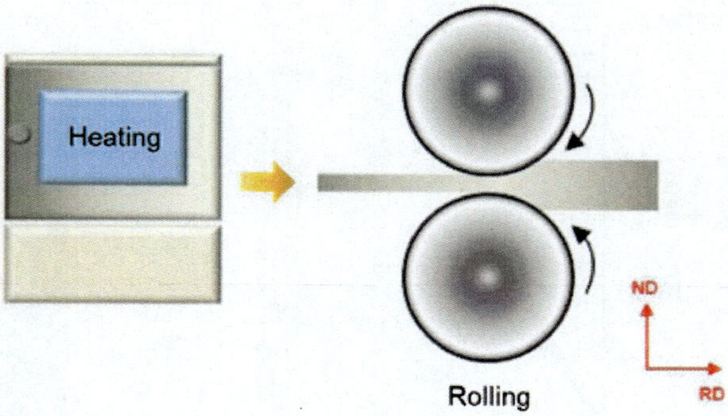

Fig. 5.8 Schematic representation of hot-rolling (annealing + rolling) process

deposits. For example, the strength and ductility of Al/B_4C composite deposits can be increased from 37 MPa and 0.3% in as-fabricated state to 131 MPa and 5.2% after hot rolling. The increment is much larger than that caused by conventional annealing (60 MPa and 1.6%) [30]. Figure 5.9 shows the microstructures of as-fabricated and hot-rolling deposits, where highly deformed particles and increased reinforcement density can be observed. By studying the microstructures and grain structures of the deposits before and after hot rolling. The strengthening mechanism is concluded as (i) refinement of matrix grains, (ii) uniform distribution of ultrafine grains of Al and re-distribution of B_4C particles in the composite, (iii) improvement in Al/B_4C interfacial bonding due to atomic diffusion across interparticle interfaces and (iv) reduction in porosity [30]. For pure metals and alloys without ceramic reinforcements, the improvement of mechanical properties is also prominent. The strength and ductility of cold sprayed A380 deposits can be improved by 320 MPa and by 5% as shown in Fig. 5.10a. The strengthening mechanism is mainly attributed to the improved interparticle bonding; the formation of nano-scaled Si particles as observed in Fig. 5.10b also contributes to the improvement [29].

Although hot rolling is a promising method for strengthening cold sprayed deposits, there is a major drawback that potentially limits its wide applications as

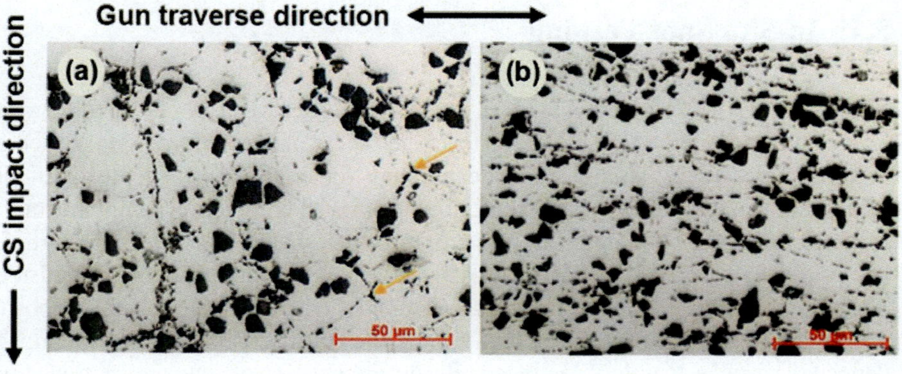

Fig. 5.9 Microstructures of **a** as-fabricated and **b** hot-rolling deposits [30]

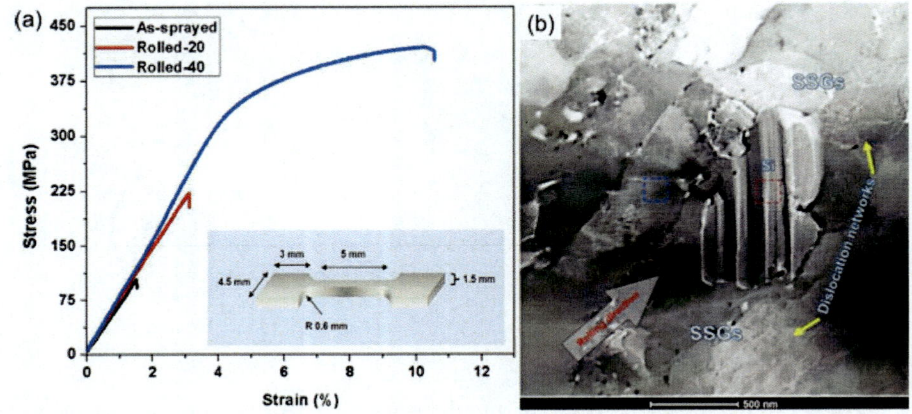

Fig. 5.10 **a** Tensile stress–strain curves of as-fabricated and hot-rolling deposits, and **b** TEM image of A380 sample after hot rolling (Rolled-40) showing the composite microstructure with nano-scaled Si fragments [29]. "Rolled-20" refers to the hot-rolling deposit with a thickness reduction of 20%; "Rolled-40" refers to the hot-rolling deposit with a thickness reduction of 40%

an additive manufacturing process. It is known that a unique advantage of additive manufacturing process over subtractive manufacturing process is the fabrication of components with net shape or near-net shape. Cold spray can not fabricate components with net shape and the as-fabricated deposits mostly require post machining in order to achieve a desirable shape, but such post machining must be minimum. Otherwise, cold spray process loses its advantage as an additive manufacturing process. Hot rolling process greatly changes the shape of as-fabricated deposits and reduces their thickness into thin plates. As such, Hot-rolling deposits likely require further extensive subtractive machining in order to produce the final ready-to-use component. One must take this fact into consideration before using hot rolling to process cold sprayed deposits.

5.4 In-Situ Shot-Peening

Shot peening is a cold working process where shot particles (i.e., round metallic, glass or ceramic) strike the surface of parts to create severe plastic deformation. Based on this principle, a new strengthening technology for cold sprayed deposits, in-situ shot-peening (also known as in-situ tamping, in-situ hammering or in-situ micro forging), was developed in recent years. In this process, large shot peening particles (i.e., typically 100–300 μm) with high hardness are added into cold spray feedstock powders, and then the mixed powders are sprayed onto substrate to form deposits. These shot-peening particles with high kinetic energy (due to its large mass) tamp the deposited particles to induce secondary plastic deformation of deposited particles and reduce defects at interparticle boundaries and thus to strengthen cold sprayed deposits

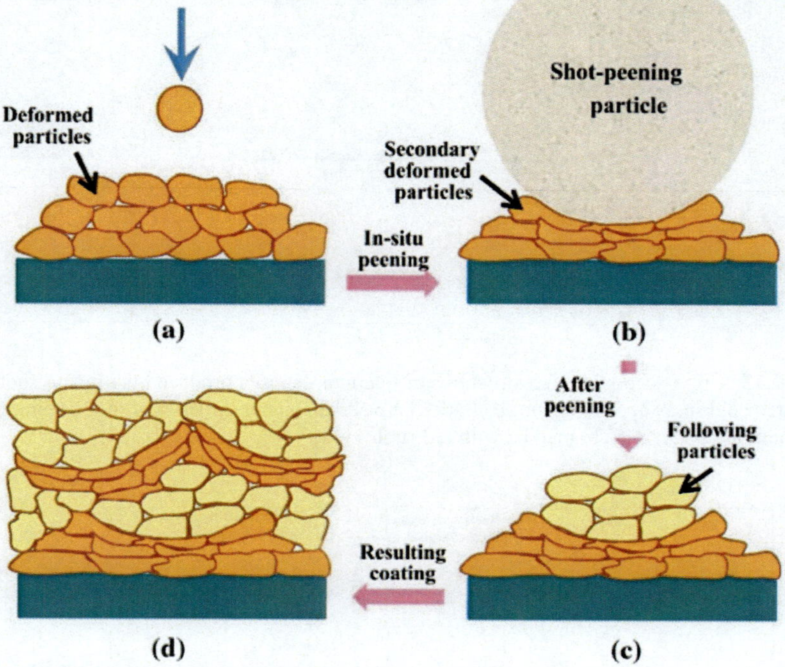

Fig. 5.11 Deposition mechanism of in-situ shot-peening assisted cold spray [32]

[31]. Meanwhile, these shot-peening particles barely deposit but rebound after impact because of their large size. Figure 5.11 schematizes the deposition mechanism of in-situ shot-peening assisted cold spray.

Cold sprayed deposit usually shows porous microstructure, particularly when the feedstock with high hardness deposition is deposited. For in-situ shot-peening assisted cold spray, the plastic deformation of deposited particles can be intensified by the shot-peening particles, which leads to improved interparticle interfaces and reduced porosity. Figure 5.12 shows the cross-sectional microstructures of titanium deposits produced using in-situ shot-peening assisted cold spray at different mixing ratios between titanium powders and shot-peening particles. As can be seen, the density of the deposits significantly increases with increasing the volume content of shot-peening particles [33]. Although the in-situ shot peening process significantly improves the density of the deposit, the microstructure sometimes exhibits inhomogeneity, which is mainly reflected by two aspects: non-tamped regions and shot-peening particle inclusion. On one hand, the regions tamped by shot-peening particles experience more severe plastic deformation than non-tamped regions, and thus the peening-affected regions have lower porosity and higher microhardness as shown in Fig. 5.13. On the other hand, the peening particles have chance to be trapped

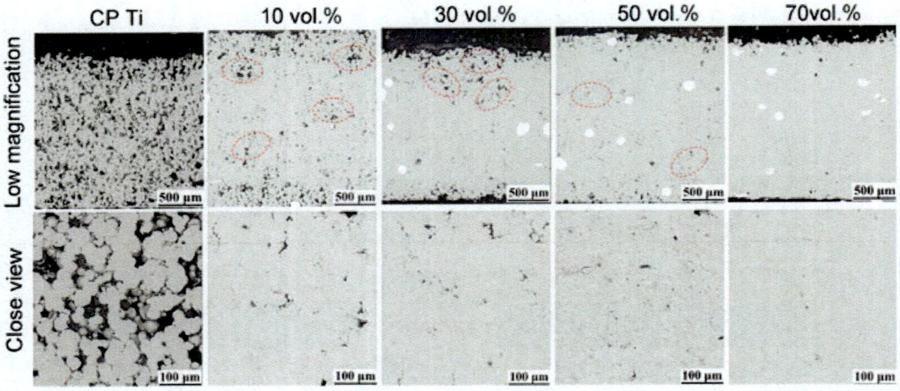

Fig. 5.12 Cross-sectional microstructures of titanium deposits produced via in-situ shot-peening assisted cold spray at different mixing ratios between titanium powders and shot-peening particles. The non-tamped zones are marked with red circles [33]

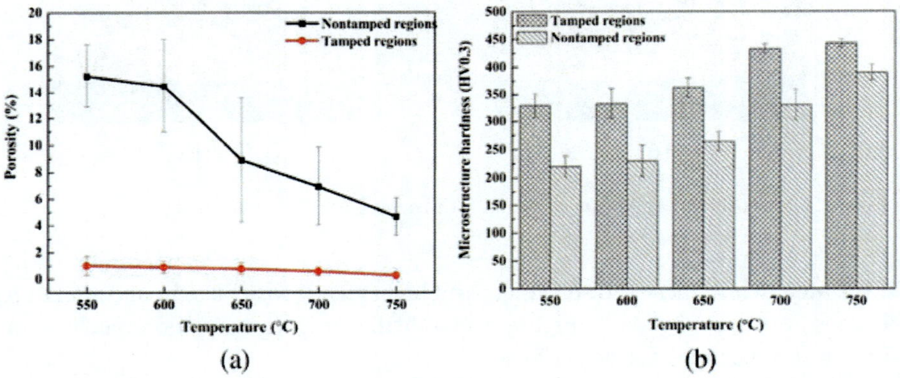

Fig. 5.13 **a** Porosity and **b** Vickers microhardness at non-tamped region and tamped region of Ti6Al4V deposits as a function of the gas temperature [32]

in the deposit and form contaminants. Such remained peening particles in the deposit can be found in Fig. 5.12.

As for the mechanical properties, Table 5.2 summarizes the detailed processing parameters and the resultant properties of the deposits fabricated via in-situ shot peening assisted cold spray. Figure 5.14 shows the stress–strain curves and microhardness of as-fabricated and heat-treated Inconel 718 deposit produced via in-situ shot-peening assisted cold spray [34]. As can be seen, the as-fabricated deposits produced with in-situ shot peening particles have higher strength than the deposits without the use of peening particles due to improved metallurgical bonding, but have lower ductility at the same time due to enhanced working hardening effect. For the same reason, the hardness of the as-fabricated deposit made via in-situ shot-peening

Table 5.2 Summary of the detailed processing parameters and the resultant properties of the deposits fabricated via in-situ shot peening assisted cold spray

Feedstock	Substrate	Gas parameters			Shot peening particles		Heat treatment	Porosity (%)	DE (%)	Mechanical properties				References
		Type	Pressure (MPa)	Temp. (°C)	Material	Cont (%)				Bonding strength (MPa)	Hardness (HV)	UTS (MPa)	EL (%)	
CP Ti (10 ~ 50 μm)	1Cr18Ni9 plate	N_2	2.8	550	1Cr13 (120 ~ 180 μm)	–	–	~ 13.8	85	–	~ 125 $HV_{0.3}$	–	–	[33]
						10		~ 4.1	–		~ 140 $HV_{0.3}$			
						30		~ 2.3	–		~ 190 $HV_{0.3}$			
						50		~ 1	–		~ 200 $HV_{0.3}$			
						70		0.3	73		~ 210 $HV_{0.3}$			
		He				–	–	1.3	89.6	–	192 $HV_{0.3}$	–	–	
Ti6Al4V (10 ~ 50 μm)	1Cr18Ni9 plate	N_2	2.8	550	1Cr13 (120 ~ 180 μm)	–		~ 15.5	81	–	~ 210 $HV_{0.3}$	–	–	
						10		~ 11.6	–		~ 235 $HV_{0.3}$			
						30		~ 5.2	–		~ 325 $HV_{0.3}$			
						50		~ 2.5	–		~ 380 $HV_{0.3}$			
						70		0.7	67		~ 415 $HV_{0.3}$			
		He	2.8	550		–		2.7	83.7	–	363 $HV_{0.3}$	–	–	
Ti6Al4V (15 ~ 30 μm)	Ti6Al4V plate	N_2	3.0	550 ~ 750	1Cr18 (125 ~ 300 μm)	70		–	64.1 ~ 74.8	26.6 ~ 36.5	–			[4]
Al6061 (20 ~ 50 μm)	AZ31B plate	N_2	3.0	350	1Cr18 (200 ~ 300 μm)	–		14.4	53.1	–	–	–	–	[35]
			2.5	300		20		10.7	~ 42.5					
						40		5.9	~ 37.5					
						60		0.4	31.2					

(continued)

Table 5.2 (continued)

Feedstock	Substrate	Gas parameters			Shot peening particles		Heat treatment	Porosity (%)	DE (%)	Mechanical properties				References
		Type	Pressure (MPa)	Temp. (°C)	Material	Cont (%)				Bonding strength (MPa)	Hardness (HV)	UTS (MPa)	EL (%)	
Ni (10 ~ 50 μm)	AZ31B plate	N_2	2.5	400	1Cr13 (150 ~ 200 μm)	40		–	–	> 65.4	–	–	–	[36]
Al (10 ~ 50 μm)	AZ31B plate	N_2	2.0	200	1Cr13 (200 ~ 300 μm)	50		0.34 ± 0.11	–	91.5 ± 2.3	64.2 ± 2.8 $HV_{0.1}$	–	–	[37]
AA2219 (10 ~ 50 μm)			2.5	300				0.23 ± 0.09		83.6 ± 2.8	118.5 ± 3.6 $HV_{0.1}$			
AA6061 (10 ~ 50 μm)								0.24 ± 0.08		87.5 ± 2.3	97.4 ± 3.8 $HV_{0.1}$			
Al (20 ~ 60 μm)	Al and Cu	N_2	3.0	330	1Cr13 (~251 μm)	–		12.4						[38]
						10		~ 9						
						30		~ 3.6						
						50		~ 1.5						
						70		0.2						
Al6061 (10 ~ 60 μm)	Al plate	N_2	2.5	300	1Cr13 (200 ~ 300 μm)	70	As-sprayed	0.12				277.6 ± 7.1	0.56 ± 0.1	[39]
							Stress relieving	0.14				298.1 ± 9.0	~ 0.9	
							Recrystallization annealing	0.37				210.8 ± 5.9	~ 9.5	
							T6	0.35				314.3 ± 7.9	~ 7.5	
IN718 (5 ~ 33 μm)	316 stainless steel plate	N_2	2.5	700	1Cr13 (~150 μm)	–	As-sprayed	5.7	23		~ 405 $HV_{0.3}$	96.4	0.12	[34]
							1200 °C for 6 h	~ 7.1			~ 380 $HV_{0.3}$	566.1	0.56	
						25	As-sprayed	1.5	34		–	–	–	
						50	As-sprayed	0.27	29.5		507 $HV_{0.3}$	463	0.48	

(continued)

Table 5.2 (continued)

Feedstock	Substrate	Gas parameters			Shot peening particles		Heat treatment	Porosity (%)	DE (%)	Mechanical properties				References
		Type	Pressure (MPa)	Temp. (°C)	Material	Cont (%)				Bonding strength (MPa)	Hardness (HV)	UTS (MPa)	EL (%)	
							1200 °C for 6 h	~ 0.4			~ 403 $HV_{0.3}$	1088.7	6.17	
						70	As sprayed	0.17	15.1		–	–	–	
Ti6Al4V (15 ~ 30 μm)	Ti6Al4V	N_2	2.5	680	1Cr13 (125 ~ 300 μm)	70	As-sprayed	7.10		~ 30		~ 95	~ 0.29	[40]
							600 °C for 2 h	6.80		~ 34		~ 115	~ 0.35	
							800 °C for 2 h	3.17		~ 54		316.46	0.54	
							1000 °C for 2 h	2.83		~ 58		–	–	
Al	LA43M Mg-Li alloy plate	N_2	3.0	280	1Cr18 (240 ~ 340 μm)	–	–	0.2	–	–	–	–	–	[41]
2219 Al alloy (5 ~ 50 μm)	FSW AA2219-T87	N_2	2.8	350	Stainless steel (150 ~ 200 μm)	–	–	2.56	–	–		–	–	[42]
						60		0.35			120 ~ 140 $HV_{0.2}$			

Temp—temperature, Cont—content, DE—deposition efficiency, UTS—ultimate tensile strength, EL—elongation

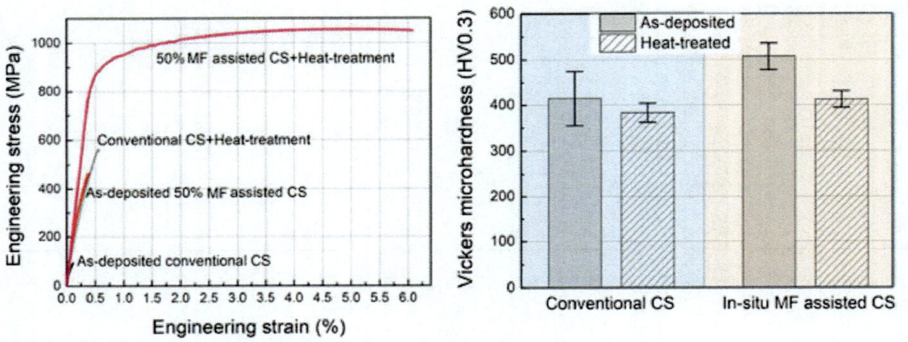

(a) (b)

Fig. 5.14 **a** Stress–strain curves and **b** microhardness of as-fabricated and heat-treated Inconel 718 deposits provided with and without in-situ shot peening [34]

assisted cold spray is much higher. It is also noted from Fig. 5.14 that both strength and ductility are significantly improved after heat treatment [34].

5.5 In-Situ Densification

In-situ shot peening assisted cold spray is indeed capable of manufacturing deposits with well-bond interparticle interfaces and less pores, which helps with the improvement of the mechanical properties and corrosion resistance of the deposits. However, it is notable that this process, as discussed above, strengthens the deposits through the impact and rebound of foreign peening particles which are different from cold sprayed powders. These foreign particles are likely to embed in the deposit and result in contamination, undesirable stress concentrations and localized chemical heterogeneities, especially when the volume content of peening particles is high [43]. In order to overcome these potential drawbacks, a new strengthening strategy, in-situ densification, was proposed. This process uses feedstock powder with wide particle size distribution and deposit the powders under low gas parameters (under low deposition efficiency) [43, 44]. Figure 5.15 shows etched microstructure (upper row) and IPF EBSD maps (lower row) of the as-fabricated Al deposits fabricated under different particle impact velocities. Under low processing parameters, a significant proportion of feedstock particles (particularly those with a large size) fail to deposit and rebound after their impact due to their insufficient impact velocities, which leads to a narrow deposition window, as shown in Fig. 5.16a. The rebounds result in a severe accumulative plastic deformation of the deposited particles and reduced porosity. When increasing processing temperature, more large particles deposit because their increasing impact velocities exceed critical velocities, seeing Fig. 5.16b. The number of rebound particles that are able to densify the deposit is reduced dramatically, resulting in a significant reduction of the total plastic deformation and leading to a

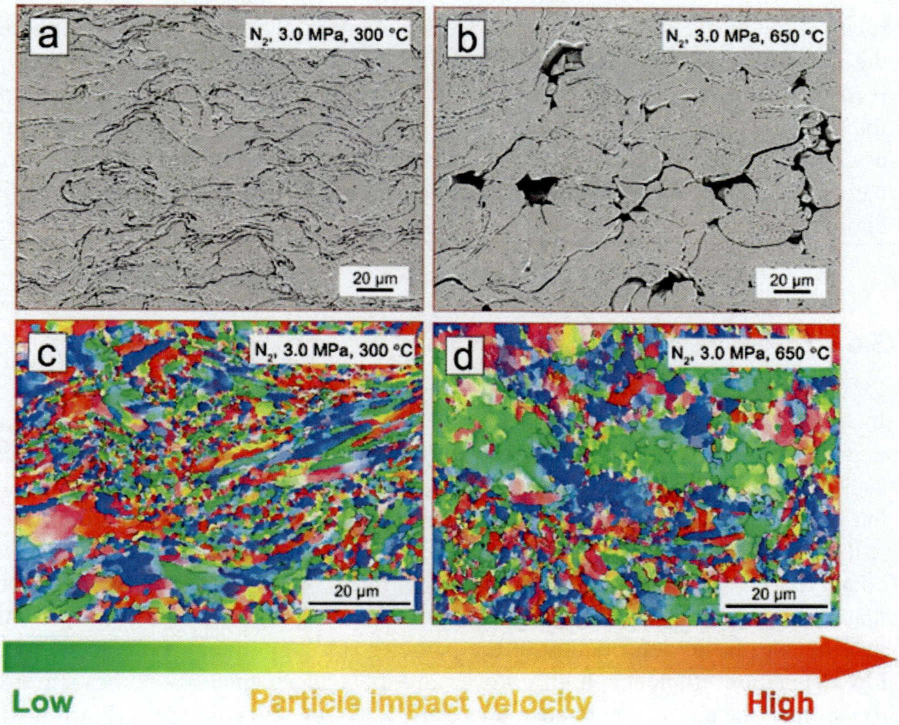

Fig. 5.15 Etched microstructure (upper row) and IPF EBSD maps (lower row) of the as-fabricated Al deposits fabricated under different particle impact velocities. The feedstock powder has a wide range size from 10 to 160 μm. [43]

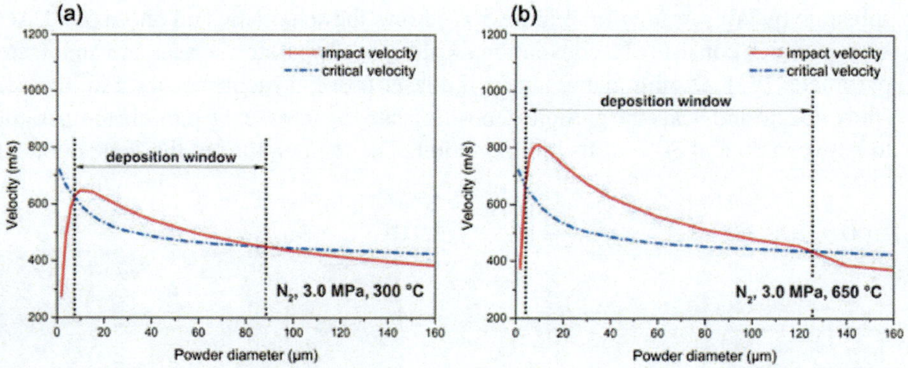

Fig. 5.16 Calculated impact and critical velocities of Al powder particles as a function of their diameters. The particles would deposit as long as their velocity is higher than the critical value. Note the significant difference in the sprayable ranges among the three different processing parameter sets used in this work [43]

relatively high porosity formation and weak interparticle interfaces [43]. Although the rebound of particles leads to a decreasing deposition efficiency and increasing feedstock consumption, the powders with wide size range are significantly cheaper than those with narrow size ranges. Moreover, compared with in-situ shot peening assisted cold spray, powder mixing procedure is no longer required, and contamination brought by shot-peening particle inclusions can be completely avoided when applying this new strategy.

5.6 Laser-Assistant Cold Spray

In Chap. 4, it has been clarified that the cohesion of conventional cold sprayed deposits predominantly relies on the interparticle metallurgical bonding. However, such metallurgical bonding does not happen over the entire interparticle interface but only occurs at the periphery of the contact interface, which limits the mechanical properties of cold sprayed deposits as compared to conventionally manufactured counterparts [45]. Particularly, when high-hardness or high strength-to-weight ratio materials such as stellite-6, tungsten and titanium are deposited via cold spray, the insufficient plastic deformation of the particles results in the deposits having very low deposition efficiency, high porosity and poor mechanical properties. In order to facilitate the deposition of such materials, laser-assisted cold spray (LASC), also known as supersonic laser deposition (SLD), was developed [46]. This hybrid manufacturing process takes the advantages of both cold spray and laser irradiation. The strengthening of the deposits is achieved through the enhance of the plastic deformation and the consequent metallurgical bonding through softening the particles and substrate by laser irradiation. Figure 5.17 shows the schematic and photo of a LACS system which consists of a conventional cold spray apparatus, a laser and an infrared pyrometer [47]. During manufacture, the laser moves synchronously with the cold spray nozzle and heats the particle deposition site to increase the particle temperature to between 30 and 80% of its melting point. The temperature of the deposition site

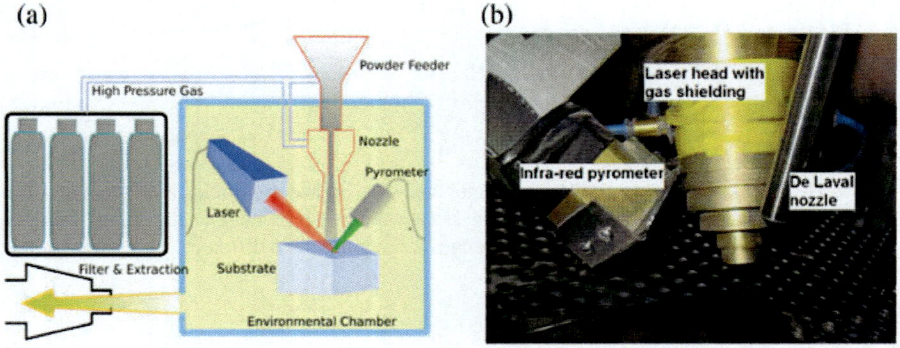

Fig. 5.17 Laser-assisted cold spray system. [47]

is measured in real-time by the pyrometer equipped on the LACS system. The heat input from laser irritation significantly softens the particle and substrate, and reduce the critical velocity, allowing the particle to deposit at lower velocity. For the same reason, the deposits fabricated via LACS demonstrate improved interparticle bonding and deposition efficiency and reduced porosity [47, 48]. Meanwhile, the laser irradiation can be well controlled to guarantee that the particle temperature does not exceed their melting temperature to prevent the possible phase transformation.

The SEM images shown in Fig. 5.18 compare the cross-sectional morphology of copper-alumina deposits between cold spray and LACS [46]. Clearly, the deposit fabricated via LACS has higher thickness (indicating higher deposition efficiency) than that made via cold spray due to the improved plastic deformation and interparticle bonding. Increasing laser irradiation of LACS will further increase the deposition efficiency [48–51], but increased irradiation will also result in softening of the deposit and thus reduced microhardness [48]. Since the deposition temperature does not cross the melting temperature of the powder materials, the deposits fabricated via LACS can largely retain their original composition and microstructure as revealed by the XRD spectra of stellite-6 powder and the corresponding LACS deposit shown in Fig. 5.19 [52]. As for the mechanical properties, the tensile strength of LACS

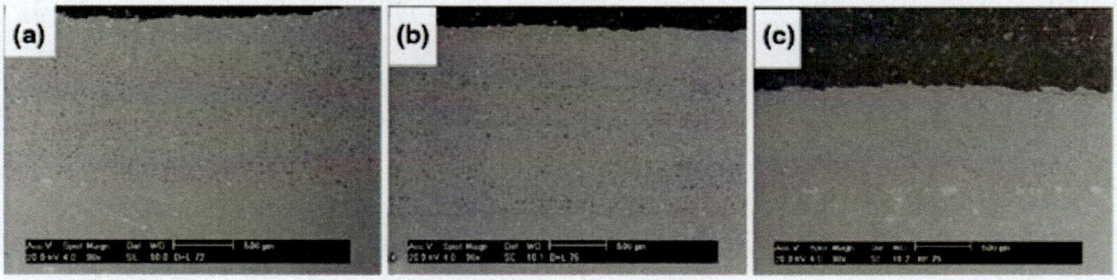

Fig. 5.18 Comparison of the cross-sectional morphology of copper-alumina deposits between cold spray and LACS. **a** 650 °C, **b** 800 °C, and **c** without the laser assistant [46]

Fig. 5.19 XRD spectra of stellite-6 powder and the corresponding LACS deposit [52]

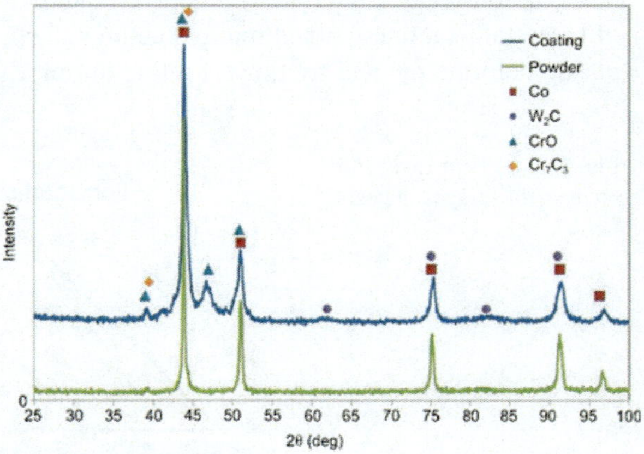

deposit (tungsten [53] and 316L stainless steel [54]) is found close to their counterparts manufactured through conventional processes. The adhesion strength between the LACS deposit and substrate also prevails [55].

Although LACS is a promising process for depositing certain materials, as compared with conventional cold spray process, LACS causes more severe oxidization of deposits due to the higher deposition temperature [55, 56]. In addition, for the materials that are easy to deposit via conventional cold spray (e.g., copper and stainless steel), LACS is less competitive from the perspective of processing cost due to additional laser apparatus and energy. LACS may also cause thermal distortion of deposit if the applied laser powder is too high. These factors must be taken into account before using LACS. Currently, LACS has been mainly applied for the deposition of materials such as titanium [47, 57], titanium alloy [58], tungsten [53], Ni60 [59], stellite-6 [48, 51, 52, 60, 61] and their composites [49, 50, 62–64].

5.7 Friction Stirring Processing

Friction stirring processing (FSP), derived from friction stir welding, uses a rotating non-consumable cylindrical tool composed of a shoulder with a pin at the tip to penetrate through the deposit and to induce local thermomechanical treatment to the deposits as schematized in Fig. 5.20 [65–75]. FSP changes the microstructure of cold sprayed deposits in the manner of particle breakdown and redistribution, homogenisation and grain refinement.

Several studies have suggested that the cold sprayed deposits processed with FSP have improved mechanical properties (i.e., tensile strength and ductility). The tensile strength and elongation of cold sprayed aluminum deposit can be increased from 60 MPa and 4.2% to 87 MPa and 60.3% [71], and for Cu60Zn40 deposit the increment is 148% in tensile strength and threefold in elongation [67]. The strengthening mechanism is attributed to the elimination of defects (e.g. porosities and non-bonded interfaces), improved interparticle bonding and the formation of recrystallized fine-grained microstructures in stirred region caused by extensive plastic deformation and frictional heating during FSP. The EBSD results shown in

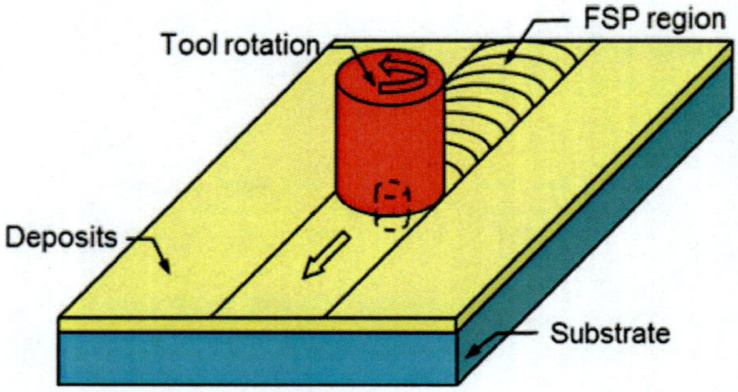

Fig. 5.20 Schematic of FSP on the cold-sprayed deposits

Fig. 5.21 clearly show the redistributed and refined grains in FSP zone. Furthermore, FSP is also capable of strengthening cold sprayed metal matrix composites. It is found that FSP broke, redistributed and refined the TiB_2 reinforcing clusters as indicated in Fig. 5.22 [74]. This fact together with refined grains and improved interparticle

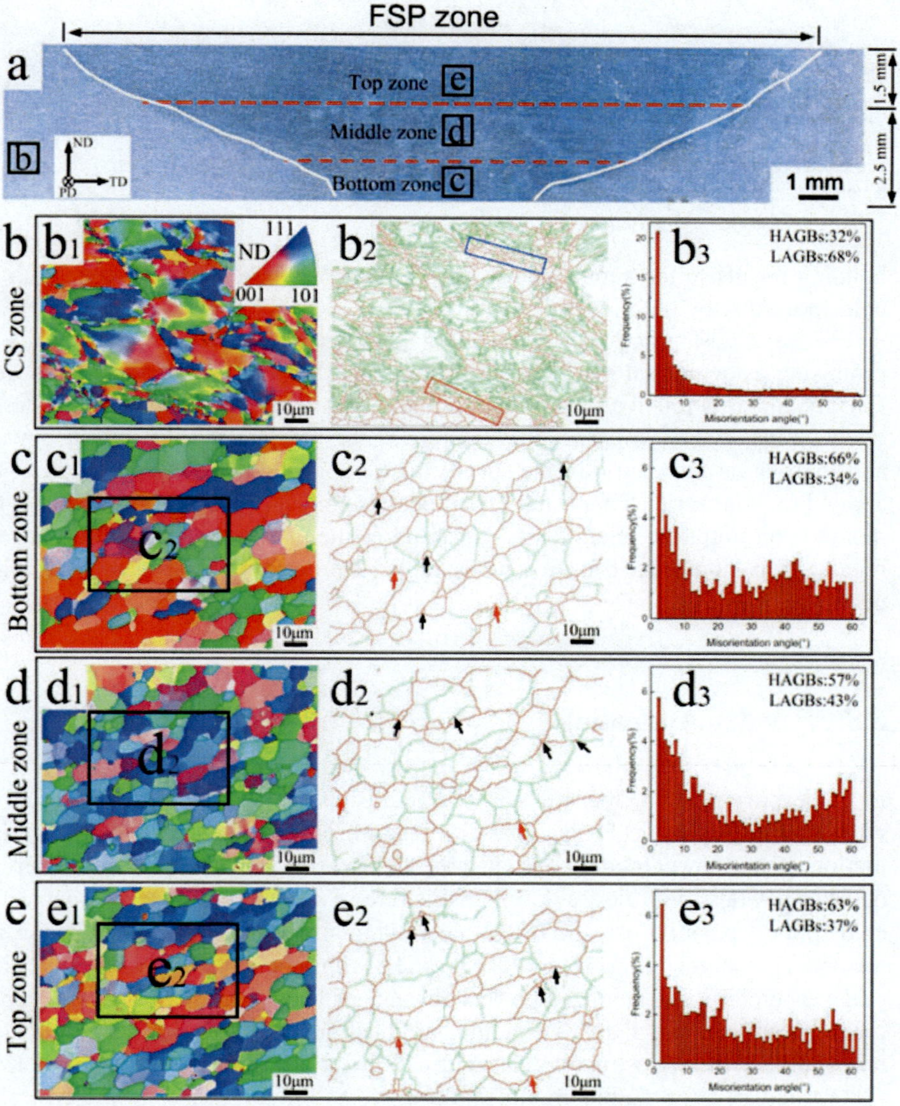

Fig. 5.21 **a** Cross-sectional macrostructure of the FSP zones; microstructure of **b** as-fabricated zone, **c** bottom, **d** middle, and **e** top zones; (b1–e1) inverse polar figure; (b2–e2) grain boundaries (the red lines represent HAGBs, while the green lines represent LAGBs); (b3–e3) distributions of misorientation angle [71]

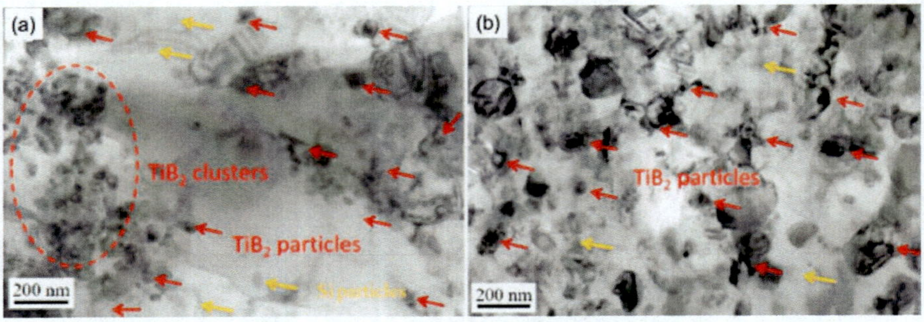

Fig. 5.22 TEM images of TiB_2 particle distributions of the TiB_2/AlSi10Mg composite deposits **a** before and **b** after FSP treatment [74]

bonding results in the simultaneous increase in tensile strength and elongation of cold sprayed AlSi10Mg-TiB_2 deposits.

FSP as a post processing method for strengthening cold sprayed deposits is promising as evidenced by the experimental data provided in literature. However, FSP treatment is relatively complicated and costly. Normally, in order activate an effective material flow during FSP, the processing parameters (e.g., tool geometry and its rotation rate and traverse speed) must be well studied before they can be applied for cold sprayed deposits. This process significantly increases the time- and money- consumption. In addition, FSP requires the part to be processed to have a regular shape or profile. Thus, it is not suitable for processing deposits with complex geometry.

5.8 Powder Annealing

In cold spray process, the intrinsic characteristics of powder feedstock such as microstructures and mechanical properties significantly influence the plastic deformation of the powders and consequently the deposition. Thus, it is important to develop powders with desirable microstructure so that they are more suitable for cold spray deposition. As one of the most commonly used powder manufacturing methods, gas atomization can produce powders with spherical shape and proper size range. Therefore, gas-atomized powders have been widely used in cold spray deposition. Despite its popularity, gas atomization has an inevitable shortcoming that is non-equilibrium microstructure caused by high solidification rate (up to 10^7 K/s) [2]. Particularly, for aluminum alloys, the powders produced in this way typically exhibit cellular/dendritic microstructure with a significantly localized segregation of solute atoms along the cell boundaries [76–83] and thus have high hardness and low ductility, which is not favorable for cold spray deposition.

Aiming to eliminate the dendritic structure and facilitate the cold spray deposition of aluminum alloy powders, solution heat treatment above the solvus temperature but below the solidus temperature has been considered as a pre-processing method to modify the microstructure and properties of as-received gas-atomized aluminum alloy powders. Figure 5.23 compares the microstructure of Al 6061powder in as-received and solution heat-treated states. Cell-like solidification microstructure with solute atoms segregated along the cell boundaries can be observed in the as-received Al 6061 powder as shown in Fig. 5.23a. The network of intermetallics is identified as the mixture of Mg-Si and Fe-Si-Al phases [80, 81]. In contrast, the heat-treated powder shown in Fig. 5.23b present a more homogeneous microstructure due to redissolution into solid solution. Similar microstructure evolution phenomenon after solution heat treatment can also be observed in Al 7075 [76, 78, 79] and Al-Cu [82] powders. Moreover, the heat treatment also leads to grain growth in the powder [77, 80, 82] and hardness reduction [77, 78]. Overall, existing studies have demonstrated that a proper heat treatment on feedstock powders can lead to the uniform distribution

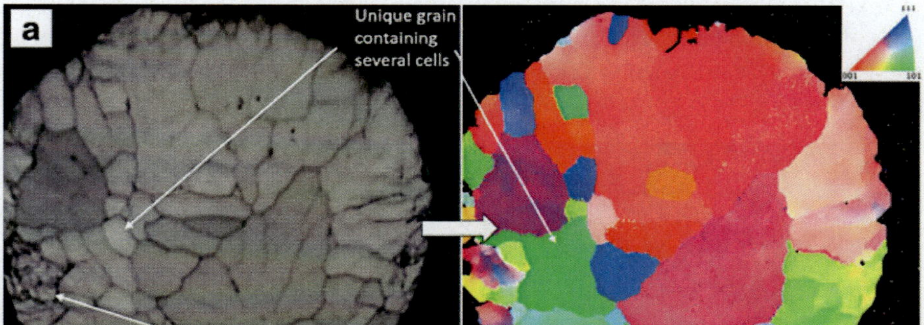

Fig. 5.23 Microstructure of Al 6061powder in its as-received and solution heat-treated states. **a** EBSD characterization of as-atomized Al6061 powder showing inhomogeneous grains with dendritic cells and **b** solution heat-treated Al6061 powder showing larger equiaxed grains [77]

of alloying elements in the powder and soften the powder, resulting in more extensive plastic deformation during deposition.

Generally, the occurrence of adiabatic shear instabilities and the formation of metal jet at the interparticle or particle–substrate interfaces are regarded as a high-quality bonding in cold spray. Solution heat-treated powders, due to their reduced hardness, perform better than as-received powders in interparticle or particle–substrate bonding. Figure 5.24a illustrates a morphology of an as-received Al 7075 particle depositing on an Al6061-T6 substrate. The particle is partially embedded into the substrate and there is little to no jetting phenomenon observed due to the high hardness and the consequent low plastic deformation upon impact, which typically indicates an unfavorable bonding [78]. In contrast, the solution heat-treated particle shown in Fig. 5.24b appears to experience a much more extensive deformation, evidenced by the higher flattening ratio and visible material jets at the splat periphery [78], which is a clear sign of "good" bond. Similar results were also reported in CuAl10Fe5Ni5 powders [84]. To quantitatively characterize the mechanical properties of deposited particle, in-situ micro-tension testing was conducted on as-received and solution heat-treated Al6061 splats by Flanagan et al. [85]. The stress–strain

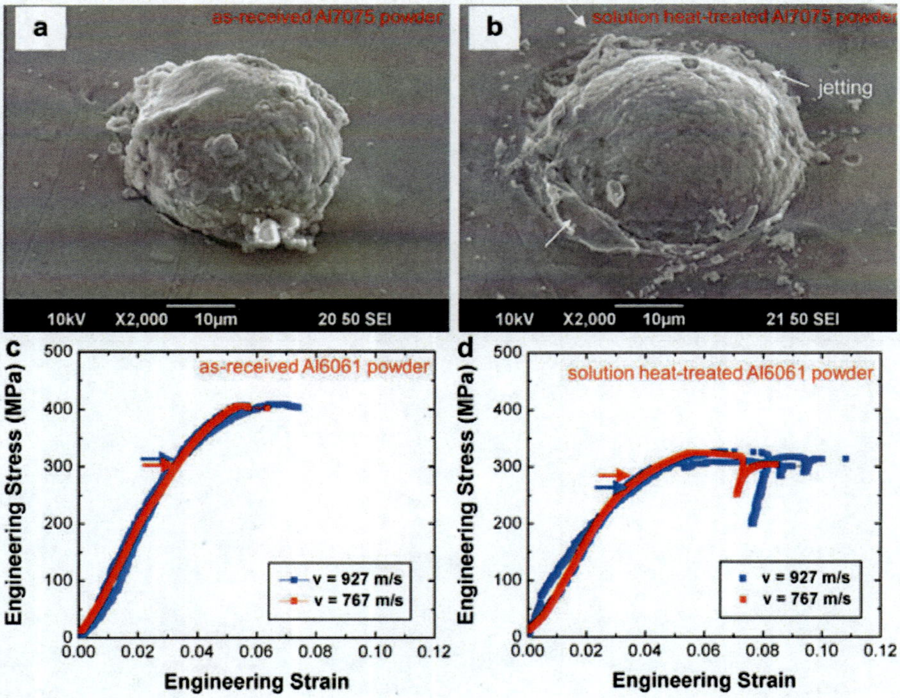

Fig. 5.24 Impact morphology of **a** an as-received and **b** an solution heat-treated Al7075 particle on a Al 6061-T6 substrate [78]; stress–strain curves obtained from micro-tension tests of **c** an as-received and **d** solution heat-treated Al6061 particle after deposition [85]

curves shown in Fig. 5.24c, d show that the as-received Al6061 splat has an elongation of 4–6% before failure, while the value is 8–10% for solution heat-treated splat. The comparison clearly demonstrates that solution heat-treated powder has better ductility than as-received powder.

Powder heat treatment helps the densification of cold sprayed deposits. Using as-received powders for cold spray results in a deposit with a porosity of $1.9 \pm 0.2\%$, while using solution heat treated powders reduces the porosity to $0.9 \pm 0.1\%$ due to reduced particle hardness [84]. Apart from the improvement on deposit density, pre-heat-treatment on feedstock powders also significantly improves the deposition efficiency of the deposit [78, 79, 82–84, 86]. This is also due to the improved powder ductility which facilitates their deposition during cold spray. This fact indicates the importance of powder heat treatment for removing the intermetallic network from the feedstock powders. Currently, the mechanical properties (e.g., yield strength, ultimate tensile strength and elongation) of cold sprayed deposits produced with heat-treated powders is still missing. Therefore, it is hard to conclude if powder heat treatment contributes to the mechanical property improvement or not. It is expected such experiments can be conducted in the near future.

References

1. Huang, R., Sone, M., Ma, W., Fukanuma, H.: The effects of heat treatment on the mechanical properties of cold-sprayed coatings. Surf. Coat. Technol. **261**, 278–288 (2015). https://doi.org/10.1016/j.surfcoat.2014.11.017
2. Ito, K., Ogawa, K.: Effects of spark-plasma sintering treatment on cold-sprayed copper coatings. J. Therm. Spray Technol. **23**, 104–113 (2014). https://doi.org/10.1007/s11666-013-0047-0
3. Koivuluoto, H., Coleman, A., Murray, K., Kearns, M., Vuoristo, P.: High pressure cold sprayed (HPCS) and low pressure cold sprayed (LPCS) coatings prepared from OFHC Cu feedstock: overview from powder characteristics to coating properties. J. Therm. Spray Technol. **21**, 1065–1075 (2012). https://doi.org/10.1007/s11666-012-9790-x
4. Winnicki, M., Małachowska, A., Baszczuk, A., Rutkowska-Gorczyca, M., Kukla, D., Lachowicz, M., Ambroziak, A.: Corrosion protection and electrical conductivity of copper coatings deposited by low-pressure cold spraying. Surf. Coat. Technol. **318**, 90–98 (2017). https://doi.org/10.1016/j.surfcoat.2016.12.101
5. Gärtner, F., Stoltenhoff, T., Voyer, J., Kreye, H., Riekehr, S., Koçak, M.: Mechanical properties of cold-sprayed and thermally sprayed copper coatings. Surf. Coat. Technol. **200**, 6770–6782 (2006). https://doi.org/10.1016/j.surfcoat.2005.10.007
6. Stoltenhoff, T., Borchers, C., Gärtner, F., Kreye, H.: Microstructures and key properties of cold-sprayed and thermally sprayed copper coatings. Surf. Coat. Technol. **200**, 4947–4960 (2006). https://doi.org/10.1016/j.surfcoat.2005.10.007
7. Kang, N., Coddet, P., Liao, H., Coddet, C.: Cold gas dynamic spraying of a novel micro-alloyed copper: microstructure, mechanical properties. J. Alloys Compd. **686**, 399–406 (2016). https://doi.org/10.1016/j.jallcom.2016.06.037
8. Borchers, C., Gärtner, F., Stoltenhoff, T., Kreye, H.: Formation of persistent dislocation loops by ultra-high strain-rate deformation during cold spraying. Acta Mater. **53**, 2991–3000 (2005). https://doi.org/10.1016/j.actamat.2005.02.048
9. Sudharshan Phani, P., Srinivasa Rao, D., Joshi, S.V., Sundararajan, G.: Effect of process parameters and heat treatments on properties of cold sprayed copper coatings. J. Therm. Spray Technol. **16**, 425–434 (2007). https://doi.org/10.1007/s11666-007-9048-1

10. Sundararajan, G., Chavan, N.M., Sivakumar, G., Sudharshan Phani, P.: Evaluation of parameters for assessment of inter-splat bond strength in cold-sprayed coatings. J. Therm. Spray Technol. **19**, 1255–1266 (2010). https://doi.org/10.1007/s11666-010-9527-7
11. Sundararajan, G., Chavan, N.M., Kumar, S.: The elastic modulus of cold spray coatings: influence of inter-splat boundary cracking. J. Therm. Spray Technol. **22**, 1348–1357 (2013). https://doi.org/10.1007/s11666-013-0034-5
12. Seo, D., Ogawa, K., Sakaguchi, K., Miyamoto, N., Tsuzuki, Y.: Parameter study influencing thermal conductivity of annealed pure copper coatings deposited by selective cold spray processes. Surf. Coat. Technol. **206**, 2316–2324 (2012). https://doi.org/10.1016/j.surfcoat.2011.10.010
13. Yin, S., Rocco, L.: Microstructure and mechanical anisotropy of additively manufactured cold spray copper deposits. Mater. Sci. Eng. A. (2018)
14. Li, W.-Y., Li, C.-J., Liao, H.: Effect of annealing treatment on the microstructure and properties of cold-sprayed Cu coating. J. Therm. Spray Technol. **15**, 206–211 (2006). https://doi.org/10.1361/105996306X108066
15. Yin, S., Cizek, J., Yan, X., Lupoi, R.: Annealing strategies for enhancing mechanical properties of additively manufactured 316L stainless steel deposited by cold spray. Surf. Coat. Technol. 1 (2019). https://doi.org/10.1016/j.surfcoat.2019.04.012
16. Zhou, H., Li, C., Yang, H., Luo, X., Yang, G., Li, W., Hussain, T., Li, C.: Pores structure change induced by heat treatment in cold-sprayed Ti6Al4V coating. J. Therm. Spray Technol. **28**, 1199–1211 (2019). https://doi.org/10.1007/s11666-019-00882-0
17. Ren, Y.Q.Q., King, P.C.C., Yang, Y.S.S., Xiao, T.Q.Q., Chu, C., Gulizia, S., Murphy, A.B.B.: Characterization of heat treatment-induced pore structure changes in cold-sprayed titanium. Mater. Charact. **132**, 69–75 (2017). https://doi.org/10.1016/j.matchar.2017.08.006
18. Ma, W., Xie, Y., Chen, C., Fukanuma, H., Wang, J., Ren, Z., Huang, R.: Microstructural and mechanical properties of high-performance Inconel 718 alloy by cold spraying. J. Alloys Compd. **792**, 456–467 (2019). https://doi.org/10.1016/j.jallcom.2019.04.045
19. Yu, H., Tieu, A.K., Lu, C., Liu, X., Liu, M., Godbole, A., Kong, C., Qin, Q.: A new insight into ductile fracture of ultrafine-grained Al-Mg alloys. Sci. Rep. **5**, 9568 (2015). https://doi.org/10.1038/srep09568
20. Kamikawa, N., Huang, X., Tsuji, N., Hansen, N.: Strengthening mechanisms in nanostructured high-purity aluminium deformed to high strain and annealed. Acta Mater. **57**, 4198–4208 (2009). https://doi.org/10.1016/j.actamat.2009.05.017
21. Chen, C., Xie, Y., Yan, X., Yin, S., Huang, R., Zhao, R., Wang, J., Ren, Z., Liu, M., Liao, H.: Effect of hot isostatic pressing (HIP) on microstructure and mechanical properties of Ti6Al4V alloy fabricated by cold spray additive manufacturing. Addit. Manuf. **27**, 595–605 (2019). https://doi.org/10.1016/j.addma.2019.03.028
22. Petrovskiy, P., Sova, A., Doubenskaia, M., Smurov, I.: Influence of hot isostatic pressing on structure and properties of titanium cold-spray deposits. Int. J. Adv. Manuf. Technol. **102**, 819–827 (2019). https://doi.org/10.1007/s00170-018-03233-5
23. Petrovskiy, P., Travyanov, A., Cheverikin, V.V., Chereshneva, A.A., Sova, A., Smurov, I.: Effect of encapsulated hot isostatic pressing on properties of Ti6Al4V deposits produced by cold spray. Int. J. Adv. Manuf. Technol. **107**, 437–449 (2020). https://doi.org/10.1007/s00170-020-05080-9
24. Qiu, X., Tariq, N.U.H., Qi, L., Wang, J.Q., Xiong, T.Y.: A hybrid approach to improve microstructure and mechanical properties of cold spray additively manufactured A380 aluminum composites. Mater. Sci. Eng. A. 772 (2020). https://doi.org/10.1016/j.msea.2019.138828
25. Zhao, Z., Tariq, N.U.H., Tang, J., Ren, Y., Liu, H., Tong, M., Yin, L., Du, H., Wang, J., Xiong, T.: Influence of annealing on the microstructure and mechanical properties of Ti/steel clad plates fabricated via cold spray additive manufacturing and hot-rolling. Mater. Sci. Eng. A. **775**, 138968 (2020). https://doi.org/10.1016/j.msea.2020.138968
26. Zhao, Z., Tariq, N.U.H., Tang, J., Jia, C., Qiu, X., Ren, Y., Liu, H., Shen, Y., Du, H., Cui, X., Wang, J., Xiong, T.: Microstructural evolutions and mechanical characteristics of Ti/steel

clad plates fabricated through cold spray additive manufacturing followed by hot-rolling and annealing. Mater. Des. **185**, 108249 (2020). https://doi.org/10.1016/j.matdes.2019.108249
27. Zhao, Z., Tang, J., Tariq, N.U.H., Liu, H., Liu, H., Ren, Y., Tong, M., Yin, L., Du, H., Wang, J., Xiong, T.: Effect of rolling temperature on microstructure and mechanical properties of Ti/steel clad plates fabricated by cold spraying and hot-rolling. Mater. Sci. Eng. A. **795**, 139982 (2020). https://doi.org/10.1016/j.msea.2020.139982
28. Tariq, N.U.H., Gyansah, L., Qiu, X., Jia, C., Awais, H. Bin, Zheng, C., Du, H., Wang, J., Xiong, T.: Achieving strength-ductility synergy in cold spray additively manufactured Al/B 4 C composites through a hybrid post-deposition treatment. J. Mater. Sci. Technol. **35**, 1053–1063 (2019). https://doi.org/10.1016/j.jmst.2018.12.022
29. Qiu, X., Tariq, N.U.H., Qi, L., Zan, Y. ning, Wang, Y. jiang, Wang, J. qiang, Du, H., Xiong, T. ying: In-situ Sip/A380 alloy nano/micro composite formation through cold spray additive manufacturing and subsequent hot rolling treatment: microstructure and mechanical properties. J. Alloys Compd. 780, 597–606 (2019). https://doi.org/10.1016/j.jallcom.2018.11.399
30. Tariq, N.H., Gyansah, L., Qiu, X., Du, H., Wang, J.Q., Feng, B., Yan, D.S., Xiong, T.Y.: Thermo-mechanical post-treatment: a strategic approach to improve microstructure and mechanical properties of cold spray additively manufactured composites. Mater. Des. **156**, 287–299 (2018). https://doi.org/10.1016/j.matdes.2018.06.062
31. Xie, Y., Chen, C., Planche, M.P., Deng, S., Huang, R., Ren, Z., Liao, H.: Strengthened peening effect on metallurgical bonding formation in cold spray additive manufacturing. J. Therm. Spray Technol. **28**, 769–779 (2019). https://doi.org/10.1007/s11666-019-00854-4
32. Zhou, H., Li, C.C., Ji, G., Fu, S., Yang, H., Luo, X., Yang, G., Li, C.C.: Local microstructure inhomogeneity and gas temperature effect in in-situ shot-peening assisted cold-sprayed Ti-6Al-4V coating. J. Alloys Compd. **766**, 694–704 (2018). https://doi.org/10.1016/j.jallcom.2018.07.009
33. Luo, X.T., Wei, Y.K., Wang, Y., Li, C.J.: Microstructure and mechanical property of Ti and Ti6Al4V prepared by an in-situ shot peening assisted cold spraying. Mater. Des. **85**, 527–533 (2015). https://doi.org/10.1016/j.matdes.2015.07.015
34. Luo, X.T., Yao, M.L., Ma, N., Takahashi, M., Li, C.J.: Deposition behavior, microstructure and mechanical properties of an in-situ micro-forging assisted cold spray enabled additively manufactured Inconel 718 alloy. Mater. Des. **155**, 384–395 (2018). https://doi.org/10.1016/j.matdes.2018.06.024
35. Wei, Y.K., Luo, X.T., Li, C.X., Li, C.J.: Optimization of in-situ shot-peening-assisted cold spraying parameters for full corrosion protection of Mg alloy by fully dense Al-based alloy coating. J. Therm. Spray Technol. **26**, 173–183 (2017). https://doi.org/10.1007/s11666-016-0492-7
36. Wei, Y.K., Li, Y.J., Zhang, Y., Luo, X.T., Li, C.J.: Corrosion resistant nickel coating with strong adhesion on AZ31B magnesium alloy prepared by an in-situ shot-peening-assisted cold spray. Corros. Sci. **138**, 105–115 (2018). https://doi.org/10.1016/j.corsci.2018.04.018
37. Wei, Y.-K.K., Luo, X.-T.T., Ge, Y., Chu, X., Huang, G.-S.S., Li, C.-J.J.: Deposition of fully dense Al-based coatings via in-situ micro-forging assisted cold spray for excellent corrosion protection of AZ31B magnesium alloy. J. Alloys Compd. **806**, 1116–1126 (2019). https://doi.org/10.1016/j.jallcom.2019.07.279
38. Fu, S.L., Li, C.X., Wei, Y.K., Luo, X.T., Yang, G.J., Li, C.J., Li, J.L.: Novel method of aluminum to copper bonding by cold spray. J. Therm. Spray Technol. **27**, 624–640 (2018). https://doi.org/10.1007/s11666-018-0707-1
39. Wei, Y.K., Luo, X.T., Chu, X., Huang, G.S., Li, C.J.: Solid-state additive manufacturing high performance aluminum alloy 6061 enabled by an in-situ micro-forging assisted cold spray. Mater. Sci. Eng. A. **776**, 139024 (2020). https://doi.org/10.1016/j.msea.2020.139024
40. Zhou, H., Li, C., Luo, X., Yang, G., Hussain, T., Li, C.: Microstructure of cross-linked high densification network and strengthening mechanism in cold-sprayed Ti-6Al-4V coating after heat treatment. J. Therm. Spray Technol. **29**, 1054–1069 (2020). https://doi.org/10.1007/s11666-020-01042-5

41. Lu, F.F., Ma, K., Li, C.X., Yasir, M., Luo, X.T., Li, C.: Enhanced corrosion resistance of cold-sprayed and shot-peened aluminum coatings on LA43M magnesium alloy. Surf. Coat. Technol. **394**, 125865 (2020). https://doi.org/10.1016/j.surfcoat.2020.125865
42. Zhang, H., Zhao, C.Y., Guo, Q.L., Yang, R.S., Liu, L.Y., Lin, S.B.: Microstructure and corrosion behavior of friction stir welded Al alloy coated by in situ shot-peening-assisted cold spray. Acta Metall. Sin. (English Letters) **33**, 172–182 (2020). https://doi.org/10.1007/s40195-019-00966-4
43. Fan, N., Cizek, J., Huang, C., Xie, X., Chlup, Z., Jenkins, R., Lupoi, R., Yin, S.: A new strategy for strengthening additively manufactured cold spray deposits through in-process densification. Addit. Manuf. 101626 (2020). https://doi.org/10.1016/j.addma.2020.101626
44. Jenkins, R., Yin, S., Aldwell, B., Meyer, M., Lupoi, R.: New insights into the in-process densification mechanism of cold spray Al coatings: low deposition efficiency induced densification. J. Mater. Sci. Technol. **35**, 427–431 (2019). https://doi.org/10.1016/j.jmst.2018.09.045
45. Chen, C., Xie, Y., Yin, S., Planche, M.-P., Deng, S., Lupoi, R., Liao, H.: Evaluation of the interfacial bonding between particles and substrate in angular cold spray. Mater. Lett. **173**, 76–79 (2016). https://doi.org/10.1016/j.matlet.2016.03.036
46. Kulmala, M., Vuoristo, P.: Influence of process conditions in laser-assisted low-pressure cold spraying. Surf. Coat. Technol. **202**, 4503–4508 (2008). https://doi.org/10.1016/j.surfcoat.2008.04.034
47. Bray, M., Cockburn, A., O'Neill, W.: The laser-assisted cold spray process and deposit characterisation. Surf. Coat. Technol. **203**, 2851–2857 (2009). https://doi.org/10.1016/j.surfcoat.2009.02.135
48. Yao, J., Li, Z., Li, B., Yang, L.: Characteristics and bonding behavior of Stellite 6 alloy coating processed with supersonic laser deposition. J. Alloys Compd. **661**, 526–534 (2016). https://doi.org/10.1016/j.jallcom.2015.11.077
49. Luo, F., Cockburn, A., Sparkes, M., Lupoi, R., Chen, Z. jun, O'Neill, W., Yao, J. hua, Liu, R.: Performance characterization of Ni60-WC coating on steel processed with supersonic laser deposition. Def. Technol. **11**, 35–47 (2015). https://doi.org/10.1016/j.dt.2014.09.003
50. Yao, J., Yang, L., Li, B., Li, Z.: Beneficial effects of laser irradiation on the deposition process of diamond/Ni60 composite coating with cold spray. Appl. Surf. Sci. **330**, 300–308 (2015). https://doi.org/10.1016/j.apsusc.2015.01.029
51. Li, B., Jin, Y., Yao, J., Li, Z., Zhang, Q., Zhang, X.: Influence of laser irradiation on deposition characteristics of cold sprayed Stellite-6 coatings. Opt. Laser Technol. **100**, 27–39 (2018). https://doi.org/10.1016/j.optlastec.2017.09.034
52. Lupoi, R., Cockburn, A., Bryan, C., Sparkes, M., Luo, F., O'Neill, W.: Hardfacing steel with nanostructured coatings of Stellite-6 by supersonic laser deposition. Light Sci. Appl. **1**, 1–6 (2012). https://doi.org/10.1038/lsa.2012.10
53. Jones, M., Cockburn, A., Lupoi, R., Sparkes, M., O'Neill, W.: Solid-state manufacturing of tungsten deposits onto molybdenum substrates with supersonic laser deposition. Mater. Lett. **134**, 295–297 (2014). https://doi.org/10.1016/j.matlet.2014.07.091
54. Gorunov, A.I., Gilmutdinov, A.K.: Investigation of coatings of austenitic steels produced by supersonic laser deposition. Opt. Laser Technol. **88**, 157–165 (2017). https://doi.org/10.1016/j.optlastec.2016.09.012
55. Li, B., Yang, L., Li, Z., Yao, J., Zhang, Q., Chen, Z., Dong, G., Wang, L.: Beneficial effects of synchronous laser irradiation on the characteristics of cold-sprayed copper coatings. J. Therm. Spray Technol. **24**, 836–847 (2015). https://doi.org/10.1007/s11666-015-0246-y
56. Barton, D.J., Bhattiprolu, V.S., Thompson, G.B., Brewer, L.N.: Laser assisted cold spray of AISI 4340 steel. Surf. Coat. Technol. **400**, 126218 (2020). https://doi.org/10.1016/j.surfcoat.2020.126218
57. Lupoi, R., Sparkes, M., Cockburn, A., O'Neill, W.: High speed titanium coatings by supersonic laser deposition. Mater. Lett. **65**, 3205–3207 (2011). https://doi.org/10.1016/j.matlet.2011.07.014
58. Birt, A.M., Champagne, V.K., Sisson, R.D., Apelian, D.: Statistically guided development of laser-assisted cold spray for microstructural control of Ti-6Al-4V. Metall. Mater. Trans. A Phys. Metall. Mater. Sci. **48**, 1931–1943 (2017). https://doi.org/10.1007/s11661-017-3970-8

59. Yao, J., Yang, L., Li, B., Li, Z.: Characteristics and performance of hard Ni60 alloy coating produced with supersonic laser deposition technique. Mater. Des. **83**, 26–35 (2015). https://doi.org/10.1016/j.matdes.2015.05.087
60. Luo, F., Cockburn, A., Lupoi, R., Sparkes, M., O'Neill, W.: Performance comparison of Stellite 6® deposited on steel using supersonic laser deposition and laser cladding. Surf. Coat. Technol. **212**, 119–127 (2012). https://doi.org/10.1016/j.surfcoat.2012.09.031
61. Luo, F., Cockburn, A., Cai, D., Sparks, M., Lu, Y.A. of S. 6® P.I. on S.S. in S.L.D.P., Ding, C., Langford, R., O'Neill, W., Yao, J., Liu, R.: Simulation analysis of Stellite 6® particle impact on steel substrate in supersonic laser deposition process. J. Therm. Spray Technol. **24**, 378–393 (2015). https://doi.org/10.1007/s11666-014-0176-0
62. Yang, L., Li, B., Yao, J., Li, Z.: Effects of diamond size on the deposition characteristic and tribological behavior of diamond/Ni60 composite coating prepared by supersonic laser deposition. Diam. Relat. Mater. **58**, 139–148 (2015). https://doi.org/10.1016/j.diamond.2015.06.014
63. Li, B., Jin, Y., Yao, J., Li, Z., Zhang, Q.: Solid-state fabrication of WCp-reinforced Stellite-6 composite coatings with supersonic laser deposition. Surf. Coat. Technol. **321**, 386–396 (2017). https://doi.org/10.1016/j.surfcoat.2017.04.062
64. Wu, L., Zhang, G., Li, B., Wang, W., Huang, X., Chen, Z., Dong, G., Zhang, Q., Yao, J.: Study on microstructure and tribological performance of diamond/cu composite coating via supersonic laser deposition. Coatings **10** (2020). https://doi.org/10.3390/coatings10030276
65. Huang, C.J., Yan, X.C., Li, W.Y., Wang, W.B., Verdy, C., Planche, M.P., Liao, H.L., Montavon, G.: Post-spray modification of cold-sprayed Ni-Ti coatings by high-temperature vacuum annealing and friction stir processing. Appl. Surf. Sci. **451**, 56–66 (2018). https://doi.org/10.1016/j.apsusc.2018.04.257
66. Yang, K., Li, W., Niu, P., Yang, X., Xu, Y.: Cold sprayed AA2024/Al_2O_3 metal matrix composites improved by friction stir processing: Microstructure characterization, mechanical performance and strengthening mechanisms. J. Alloys Compd. **736**, 115–123 (2018). https://doi.org/10.1016/j.jallcom.2017.11.132
67. Huang, C., Li, W., Feng, Y., Planche, M.P., Liao, H., Montavon, G.: Microstructural evolution and mechanical properties enhancement of a cold-sprayed Cu[sbnd]Zn alloy coating with friction stir processing. Mater. Charact. **125**, 76–82 (2017). https://doi.org/10.1016/j.matchar.2017.01.027
68. Yang, K., Li, W., Xu, Y., Yang, X.: Using friction stir processing to augment corrosion resistance of cold sprayed AA2024/Al_2O_3 composite coatings. J. Alloys Compd. **774**, 1223–1232 (2019). https://doi.org/10.1016/j.jallcom.2018.09.386
69. Yang, K., Li, W., Huang, C., Yang, X., Xu, Y.: Optimization of cold-sprayed AA2024/Al_2O_3 metal matrix composites via friction stir processing: effect of rotation speeds. J. Mater. Sci. Technol. **34**, 2167–2177 (2018). https://doi.org/10.1016/j.jmst.2018.03.016
70. Huang, C., Li, W., Zhang, Z., Fu, M., Planche, M.P., Liao, H., Montavon, G.: Modification of a cold sprayed SiCp/Al5056 composite coating by friction stir processing. Surf. Coat. Technol. **296**, 69–75 (2016). https://doi.org/10.1016/j.surfcoat.2016.04.016
71. Wang, W., Han, P., Wang, Y., Zhang, T., Peng, P., Qiao, K., Wang, Z., Liu, Z., Wang, K.: High-performance bulk pure Al prepared through cold spray-friction stir processing composite additive manufacturing. J. Mater. Res. Technol. **9**, 9073–9079 (2020). https://doi.org/10.1016/j.jmrt.2020.06.034
72. Khodabakhshi, F., Marzbanrad, B., Shah, L.H., Jahed, H., Gerlich, A.P.: Friction-stir processing of a cold sprayed AA7075 coating layer on the AZ31B substrate: structural homogeneity, microstructures and hardness. Surf. Coat. Technol. **331**, 116–128 (2017). https://doi.org/10.1016/j.surfcoat.2017.10.060
73. Khodabakhshi, F., Marzbanrad, B., Jahed, H., Gerlich, A.P.: Interfacial bonding mechanisms between aluminum and titanium during cold gas spraying followed by friction-stir modification. Appl. Surf. Sci. **462**, 739–752 (2018). https://doi.org/10.1016/j.apsusc.2018.08.156
74. Xie, X., Chen, C., Chen, Z., Wang, W., Yin, S., Ji, G., Liao, H.: Achieving simultaneously improved tensile strength and ductility of a nano-TiB_2/AlSi10Mg composite produced by cold spray additive manufacturing. Compos. Part B Eng. **202**, 108404 (2020). https://doi.org/10.1016/j.compositesb.2020.108404

75. Wang, Y., Huang, Y., Meng, X., Wan, L., Feng, J.: Microstructural evolution and mechanical properties of MgZnYZr alloy during friction stir processing. J. Alloys Compd. **696**, 875–883 (2017). https://doi.org/10.1016/j.jallcom.2016.12.068
76. Sabard, A., McNutt, P., Begg, H., Hussain, T.: Cold spray deposition of solution heat treated, artificially aged and naturally aged Al 7075 powder. Surf. Coat. Technol. **385**, 125367 (2020). https://doi.org/10.1016/j.surfcoat.2020.125367
77. Sabard, A., Hussain, T.: Inter-particle bonding in cold spray deposition of a gas-atomised and a solution heat-treated Al 6061 powder. J. Mater. Sci. **54**, 12061–12078 (2019). https://doi.org/10.1007/s10853-019-03736-w
78. Sabard, A., de Villiers Lovelock, H.L., Hussain, T.: Microstructural evolution in solution heat treatment of gas-atomized Al alloy (7075) powder for cold spray. J. Therm. Spray Technol. **27**, 145–158 (2017). https://doi.org/10.1007/s11666-017-0662-2
79. Story, W.A., Brewer, L.N.: Heat treatment of gas-atomized powders for cold spray deposition. Metall. Mater. Trans. A Phys. Metall. Mater. Sci. **49**, 446–449 (2018). https://doi.org/10.1007/s11661-017-4428-8
80. Bedard, B.A., Flanagan, T.J., Ernst, A.T., Nardi, A., Dongare, A.M., Brody, H.D., Champagne, V.K., Lee, S.-W., Aindow, M.: Microstructure and micromechanical response in gas-atomized Al 6061 alloy powder and cold-sprayed splats. J. Therm. Spray Technol. (2018). https://doi.org/10.1007/s11666-018-0785-0
81. Tsaknopoulos, K., Walde, C. Champagne, V. Jr., Cote, D.: Gas-atomized Al 6061 powder: phase identification and evolution during thermal treatment. Jom (2018). https://doi.org/10.1007/s11837-018-3175-7
82. Liu, T., Story, W.A., Brewer, L.N.: Effect of heat treatment on the Al-Cu feedstock powders for cold spray deposition. Metall. Mater. Trans. A Phys. Metall. Mater. Sci. **50**, 3373–3387 (2019). https://doi.org/10.1007/s11661-019-05230-z
83. Rokni, M.R., Nardi, A.T., Champagne, V.K., Nutt, S.R.: Effects of preprocessing on multi-direction properties of aluminum alloy cold-spray deposits. J. Therm. Spray Technol. **27**, 818–826 (2018). https://doi.org/10.1007/s11666-018-0723-1
84. Krebs, S., Gärtner, F., Klassen, T.: Cold spraying of Cu-Al-Bronze for cavitation protection in marine environments. J. Therm. Spray Technol. **24**, 126–135 (2014). https://doi.org/10.1007/s11666-014-0161-7
85. Flanagan, T.J., Bedard, B.A., Dongare, A.M., Brody, H.D., Nardi, A., Champagne, V.K., Aindow, M., Lee, S.W.: Mechanical properties of supersonic-impacted Al6061 powder particles. Scr. Mater. **171**, 52–56 (2019). https://doi.org/10.1016/j.scriptamat.2019.06.024
86. Ko, K.H., Choi, J.O., Lee, H.: Pretreatment effect of Cu feedstock on cold-sprayed coatings. J. Mater. Process. Technol. **214**, 1536–1545 (2014). https://doi.org/10.1016/j.jmatprotec.2014.02.020

Chapter 6
Cold Sprayed Metal Matrix Composites

Abstract Metal matrix composites (MMCs) are composite materials for which reinforcing particles and/or flakes (e.g., ceramic, diamond, 2D materials) are incorporated into metal or alloy matrix so that the new material can take the advantages of both and to gain new properties. CSAM has demonstrated superiority over conventional fusion-based AM technologies for fabricating MMCs due to its short production time, unlimited product size, high flexibility and suitability for damaged component repair [1, 2]. Over the years, extensive studies have been conducted to investigate the deposition mechanisms, microstructures and properties of cold sprayed MMC deposits. To date, various starting powders have been applied as the feedstocks for the cold spray deposition of MMCs. Typically, the MMC deposits fabricated using different feedstocks are governed by different deposition mechanisms and exhibit different microstructures. This chapter focuses on the MMC deposits made via CSAM, from feedstock preparation, deposition mechanism, microstructure to mechanical properties.

6.1 Composite Feedstock

The feedstock used for the fabrication of MMC deposits is mostly prepared before cold spraying. Common methods for the preparation of the starting feedstock include mechanical blending, ball-milling, agglomerating and sintering, and cladding. Each method has its unique advantages and disadvantages. As a summarization, Table 6.1 lists the characteristics of the aforementioned feedstock preparation methods. A detailed description of each method is provided in the following sub-sections.

6.1.1 Mechanical Blending

Mechanical blending is the most commonly used feedstock preparation method due to its simplicity and low cost. With this method, metal or alloy powders are mixed

S. Yin and R. Lupoi, *Cold Spray Additive Manufacturing*, Springer Tracts in Additive Manufacturing, https://doi.org/10.1007/978-3-030-73367-4_6

Table 6.1 Summarization of feedstock preparation methods

Method	Merit	Downside
Mechanical blending	(1) Wide-range selection of powders (2) Simplicity in feedstock preparation (3) Low cost	(1) Low deposition efficiency of reinforcing particles (2) Fragmentation of brittle reinforcing particles
Ball-milling	(1) Homogeneous distribution of reinforcing particles in deposit (2) Capability to produce nanosized reinforcing particles (3) Metallurgical bonding between interparticles	(1) Irregular powder feedstock (2) Increased powder hardness (3) Complicated feedstock preparation process
Agglomerating and sintering	(1) Spherical particle shape (2) High contents of reinforcing phase in deposit	(1) Limited selection of powder feedstock, mainly WC–Co and Cr_3C_2-NiCr (2) Low deposition efficiency (3) Complicated feedstock preparation process (4) High production cost
Cladding	(1) High deposition efficiency of reinforcing phase (2) Metallurgical bonding between interparticles (3) Prevent the fragmentation of brittle reinforcing phase	(1) Limited selections of powder feedstock (2) Complicated feedstock preparation process (3) High production cost

with reinforcing powders at a certain ratio in a container and the mixed powders are then blended mechanically to form the starting feedstock. An example of mechanically blended feedstock used for cold spray deposition is shown in Fig. 6.1 [3]. Although its popularity, mechanical blending also has some inevitable downsides. The absence of matrix-reinforcement metallurgical bonding leads to a higher likelihood of rebound of the reinforcing particles, resulting in a significant reduction of the reinforcement fraction in the deposits as compared to in the starting feedstock. Therefore, it is difficult to produce MMC deposits with a high fraction of reinforcing phase using mechanically blended feedstock [4–6]. Furthermore, the high particle impact velocities frequently result in the fracture of brittle reinforcing particles [7–9], which is an undesirable effect for applications which require high electrical and thermal conductivity [10]. In recent years, an alternative mechanical blending process was developed and applied for fabricating MMC deposits. In this process, two powder feeders are used for delivering the metal and reinforcing powders into the cold spray nozzle separately [8, 11]. The two streams of powders mix in the pre-chamber of the nozzle and then accelerate towards the substrate. An unique advantage of the alternative method is the in-situ metal-reinforcement ratio control through adjusting

Fig. 6.1 Morphology of mechanically blended Al-Al_2O_3 feedstock for cold spray deposition

the feed rate of each powder feeder, which potentially enables the manufacture of functionally graded materials.

6.1.2 Ball-Milling

Ball milling, also known as mechanical alloying, is an effective and efficient method to prepare composite powder feedstock for cold spray. It is a process where mixed powders are placed in a chamber and subjected to high-energy collision of balls to induce composite powders. By adjusting milling speed, milling time, ball-to-powder ratio and metal-to-reinforcement ratio, the composition, morphology and microstructure of the prepared composite powders can be well controlled. Figure 6.2 shows the surface morphology, cross-section and TEM image of ball-milled NiCrAl-BN composite powders used for cold spray deposition [12]. As can be seen, the ball-milled powders have non-spherical shape and the nanosized BN reinforcement uniformly distributed within NiCrAl matrix.

Ball milling is mostly used for preparing the feedstock with nanostructured reinforcement (e.g., graphene, CNT, BN). It also has some problems. The alloying process increases the strength of the metal particles, thereby lowering deposition efficiency. Further, the alloying process changes the powder morphology into an irregular shape that may cause powder feeding issues due to the inferior flowability or produce higher coating porosity. Finally, the brittle reinforcing particles often break into fragments during the alloying process. Needless to say, these effects reduce the performance

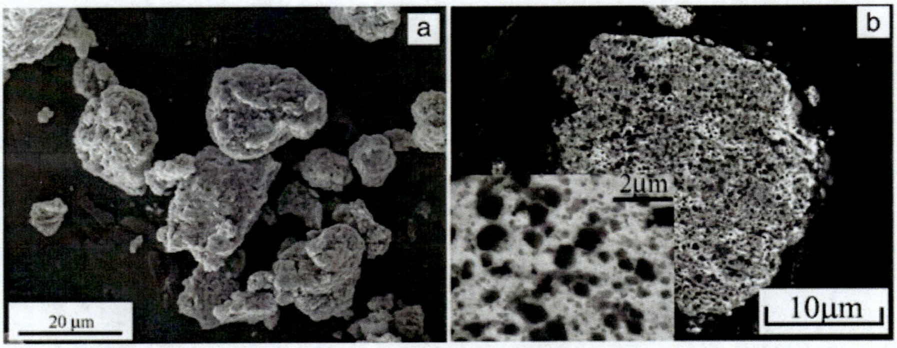

Fig. 6.2 Surface morphology, cross-section and TEM image of ball-milled NiCrAl-BN composite powders used for cold spray deposition [12]

of the deposition process and the final coating quality (i.e., decreased deposition efficiency, increased porosity, lower cohesion strength) [13–15].

6.1.3 Cladding

As addressed above, mechanical blending and ball milling both have negative downsides which may lead to undesirable performance of the MMC deposits. In an attempt to resolve this problem, metal-cladded powders composed of an inner reinforcing phase and a ductile metallic shell were applied for cold spray deposition in some studies [16–21]. Most of the metal-cladded composite powders used for cold spray were prepared by electroplating and electroless plating. Figure 6.3 shows the morphology and cross-section of copper-nickel-coated diamond powders made via electroless plating [19, 22]. The outside metal layer with a thickness of a few microns can be clearly observed, and the cladded powders demonstrate a similar shape as the

Fig. 6.3 Morphology and cross-section of copper-nickel-coated diamond powders made via electroless plating [19, 22]

diamond core. As compared to uncoated reinforcing powders, the cladded particles are more easily deposited and the retained diamond content in the deposit is higher due to the occurrence of metallurgical bonding between the shells of the cladded powders [22]. The metal shell also somewhat prevents the fracture of the brittle reinforcing core upon impact. Although the cladded powders are promising for the deposition of cold sprayed MMCs, the cladded composite powder feedstock is much more costly than mechanically blended and ball-milled feedstock, which may limit its wide application in general industry. The cladded powder can also be mechanically blended with pure metal powders to form starting feedstock.

6.1.4 Agglomerating and Sintering

Agglomerating and sintering method is typically used for fabricating cermet powders such as WC–Co and Cr_3C_2-NiCr. For preparing the cermet powders, metal powders (typically 1–2 μm in size) and reinforcing powders (typically less than 5 μm) are blended in the first place. The mixed powder is dispersed in water along with a suitable polymer binder. This dispersion is atomized into "green" powders through spray drying and then heated to remove the polymer binder and to induce sintering. Eventually, agglomerated powders are produced. The agglomerated powders usually present a spherical morphology with high volume fraction of reinforcing phase. Figure 6.4 shows the morphology and cross-section of WC-17Co powders used for cold spray deposition [23]. Deposits using sintered feedstock typically results in high fraction of reinforcing phase in the deposits, but the deposition efficiency of sintered particles is relatively low [24]. Same as the cladded powder, the manufacturing cost of sintered powder is relatively high and thus only limited range of sintered powders has been developed today.

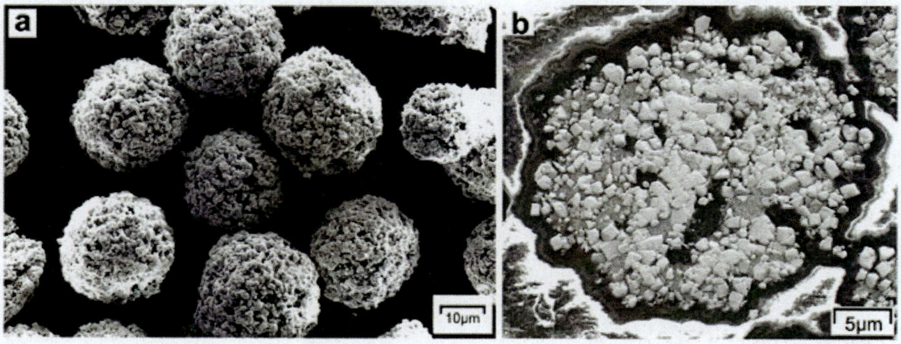

Fig. 6.4 Morphology and cross-section of WC-17Co powders used for cold spray deposition [23]

6.2 Deposition Mechanism

6.2.1 *Deposition Mechanism of Mechanically Blended Feedstock*

The deposition mechanism of cold sprayed MMCs is different from that of cold sprayed pure metals due to the introduction of additional reinforcing phases (typically ceramics) in the feedstock. For mechanically blended feedstock, the cohesion of metal particles typically follows the metal-to-metal bond mechanism while reinforcing particles deposits through penetration and interlocking into the metal phase without any chemical reason with the matrix phase [25]. The TEM images showing the interface between Al matrix and diamond reinforcement in a Al-diamond composite and the EDX line scan across the interface as shown in Fig. 6.5 confirms the lack of reaction [22].

During the co-deposition process, the hard reinforcing phase also helps to reduce the critical velocity of metal particles and promotes their deposition through "activate" the metal surface by cleaning the oxides [26, 27]. Because of this, the deposition efficiency of cold sprayed MMC deposits is higher than that of pure metal deposits under the same processing parameters as evidenced in Fig. 6.6 [5, 28, 29]. However, studies also addressed that the ceramic-ceramic collision is dominant when the fraction of reinforcing phase is too high. In this case, most of the ceramic particles may rebound after impact, which results in a reduction in deposition efficiency as compared to that of pure metal deposits (Fig. 6.6) [5, 28, 29].

A recent work from Yin et al. added some new findings into the deposition mechanism of cold sprayed MMC deposits [30]. They found that most of the reinforcing particles in the MMC deposits fracture during the deposition process with very few intact particles retained, and there exists a critical velocity to determine the transition of reinforcing particles from rebound regime to fragmentation regime. Figure 6.7 shows the in-situ observations of two diamond particles impacting onto the copper substrate at rebound regime and fragmentation regime, i.e., 190 and 335 m/s [30]. Spraying at fragmentation regime led to much more diamond content in the deposit

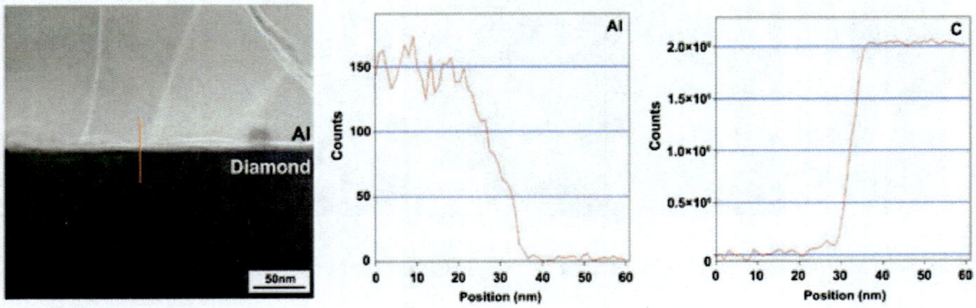

Fig. 6.5 TEM images at the interface between Al matrix and diamond reinforcement in a Al-diamond composite and the EDX line scan across the interface [22]

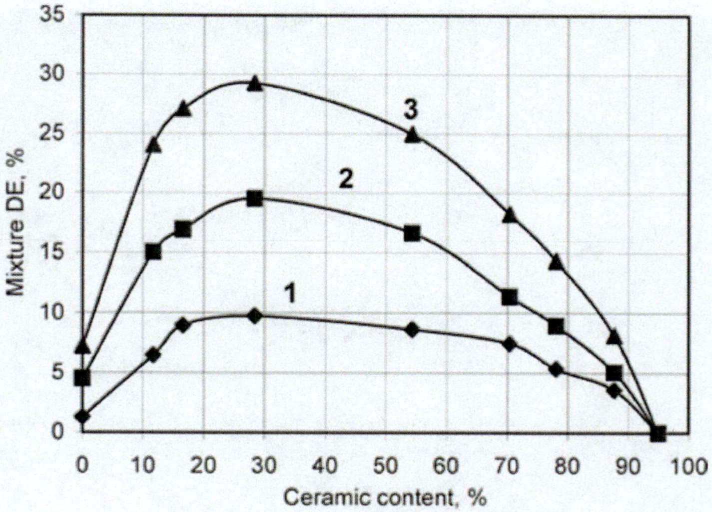

Fig. 6.6 Deposition efficiency of cold sprayed Al-Al_2O_3 MMC deposits fabricated at various gas temperatures [28]

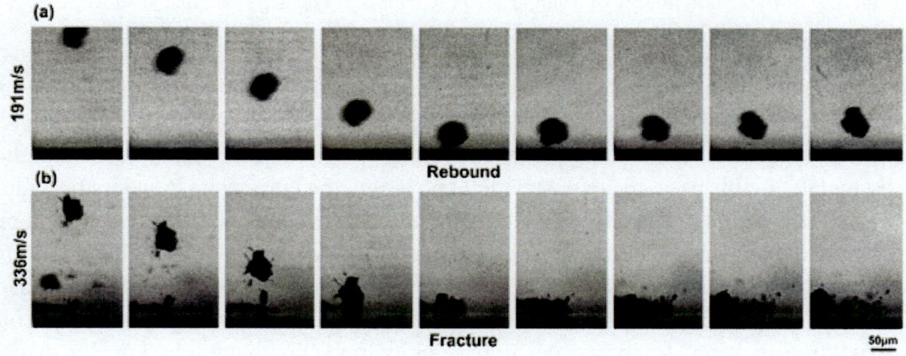

Fig. 6.7 In-situ observations of two diamond particles impacting onto the copper substrate at rebound regime and fragmentation regime, i.e., 190 and 335 m/s [30]

than spraying at rebound regime. This work demonstrates that the build-up MMC deposits during CSAM largely relies on the fragmentation of the reinforcing particles.

6.2.2 *Deposition Mechanism of Ball-Milled and Sintered Feedstock*

Hard reinforcing particles in cold sprayed MMC coatings do not undergo plastic deformation during the deposition process. Therefore, these hard particles do not

Fig. 6.8 Surface morphology of a single WC-17Co particle deposition onto a Ni substrate [23]

have metallic bonding with the soft matrix phase, tending to rebound from the coating under the effect of large bulk modulus [31, 32]. However, for ball-milled and sintered powders, the hard reinforcing particles are incorporated in or consolidated together by metals. In this case, the exposed metal phases in the composite powders provide a possibility for the occurrence of metallic bonding between the ball-milled or sintered composite powders. Therefore, it is likely that the deposition for ball-milled and sintered feedstock relies on the metallurgical bonding between the metal phases in different powders. In addition, for sintered particles, as most of them are porous structures, they may fracture upon high-velocity impact with the substrate due to high kinetic energy. An evidence of such fracture is provided in Fig. 6.8 where a single WC-17Co particle severely fractured upon impact with a Ni substrate [23]. Such fracture may cause the loss of the deposited materials and reduce their deposition efficiency.

6.2.3 Deposition Mechanism of Cladded Feedstock

As discussed above, non-metallic reinforcing particles in cold sprayed MMC deposits only mechanically embed into soft metal matrix. Importantly, they do so without triggering any metallurgical reaction because the non-metallic reinforcing phase naturally cannot initiate a metallurgical reaction with the metal matrix [25]. However, for metal-cladded reinforcing particles, an entirely different mechanism dominates the

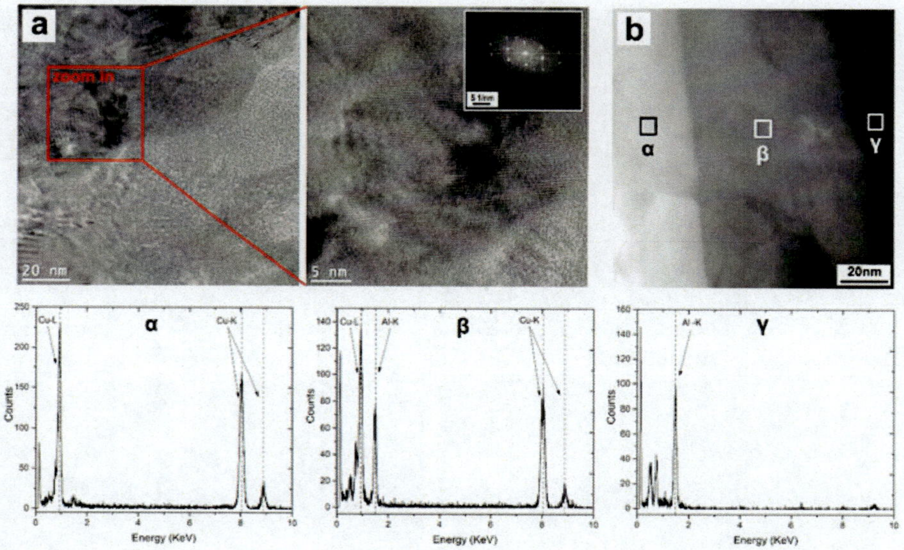

Fig. 6.9 HRTEM images and the corresponding FFT pattern and EDX analysis of the interface between the Cu shell coated on a diamond core surface and Al matrix [22]

deposition. The reinforcing particle has a metallic shell which inherently facilitates an intimate metal-to-metal contact and potentially enables a metallurgical bonding with the metal matrix [33]. Upon the high velocity impact of a metal-cladded reinforcing particle onto metal matrix, the oxide films on the surfaces of the metal shell and metal matrix are broken. The metal shell and the metal matrix both experiences significant plastic deformation and localized temperature rise. This results in the formation of outward metal jet which helps to clean the contact interface by extruding the fractured oxides outwards, providing an intimate metal-to-metal contact and thus metallurgical bonding [34, 35]. This process is actually analogous to what has been found in conventional pure metal coatings [33]. As an evidence, Fig. 6.9 shows the HRTEM images and the corresponding FFT pattern and EDX analysis of the interface between the Cu shell coated on a diamond core surface and Al matrix [22]. The formation of $CuAl_2$ intermetallic phase indicates the formation of metallurgical bonding between the Cu shell and the Al matrix [36, 37]. Due to the occurrence of such metallurgical bonding, metal-cladded reinforcing particles are much easier to deposit than uncoated particles [22].

6.3 Microstructures and Mechanical Properties

Figure 6.10 shows the typical cross-sectional microstructure of cold sprayed MMC deposits fabricated using mechanically blended, ball-milled, sintered and cladded feedstock [4, 22, 23, 38]. As can be seen, all deposits demonstrate high density and

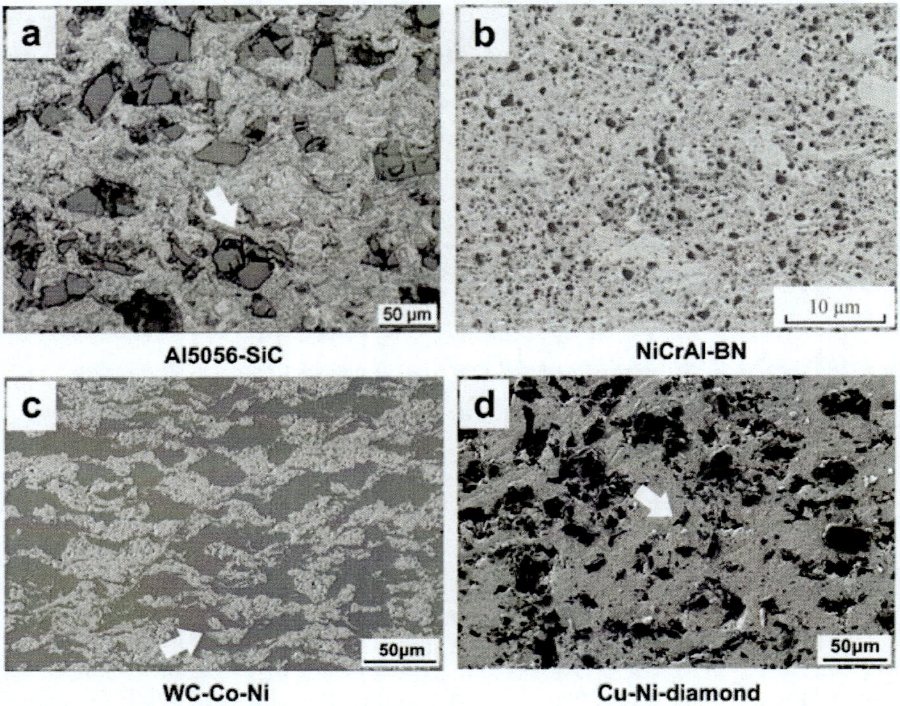

Fig. 6.10 Typical cross-sectional microstructure of cold sprayed MMC deposits fabricated using **a** mechanically blended, **b** ball-milled, **c** sintered and **d** cladded feedstock [4, 22, 23, 38]

compact structure with few to no inner pores. The metal phases have intimate contact with the reinforcing phases in the deposits. For the mechanically blended, sintered and cladded feedstock, fracture of reinforcing phases can be noticed as marked by the white arrows. For mechanically blended feedstock, the original volume ratio of reinforcing phase to metal phase in the feedstock may change after deposition due to their different deposition efficiencies. Typically, more reinforcing powders will be lost than metal powders and thus the volume ratio will reduce after deposition. The amount of the reduction is related to gas processing parameters, metal and reinforcing particle conditions (e.g., shape, size and type). For the deposit made using ball-milled feedstock, the size of BN reinforcing particles is much smaller as compared to the other three cases [38]. These reinforcing particles are pre-incorporated in the metal powders before spraying and later well retained in the deposit after deposition. For sintered WC-17Co feedstock, severe plastic deformation can be noticed as evidenced by the flattened splats due to the existence of metal phase as binders [23]. Note that the deposit shown in Fig. 10c was not fabricated using pure sintered feedstock but using a mixture of sintered WC-17Co and Ni powders. Pure WC-17Co deposit can also be produced via cold spray but with very low deposition efficiency [24]. For the cladded feedstock, the deposit has high volume fraction of diamond reinforcement. The outer metal shell mostly surrounding the reinforcing particle core can be found

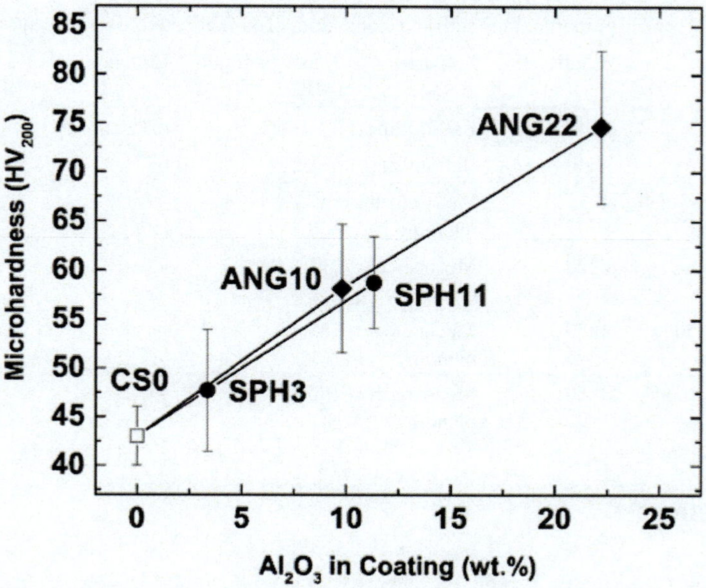

Fig. 6.11 Microhardness of Al-Al_2O_3 deposits as a function of the weight content of Al_2O_3 in the deposit. CS0: no Al_2O_3 addition, SPH: spherical powders, ANG: angular powders [39]

in the deposit; these metal shells can somewhat prevent the fracture of the brittle particle and improve the deposit performance accordingly.

Figure 6.11 shows the microhardness of Al-Al_2O_3 deposits as a function of the Al_2O_3 weight ratio in the deposit [39]. As can be seen, the microhardness increases with increasing the contents of Al_2O_3. This is mainly due to the enhanced work hardening effect of Al matrix, increased amount of hard reinforcing phase, and uniformly dispersed ceramic particles which restricts the deformation of soft Al matrix [40]. Such changing trend is almost seen in all cold sprayed MMC deposits. In recent years, as cold spray has been developing to an additive manufacturing process, tensile properties of cold sprayed MMC deposits have been studied in some studies. Table 6.2 summarizes the typical tensile properties (tensile strength and elongation) of as-fabricated cold sprayed MMC deposits. All deposits demonstrate extremely low elongation, which is normal for cold spray. Same as cold sprayed pure metal deposits, by applying proper post heat treatments (e.g., annealing [41], hot rolling [6, 42] and friction stirring [43]), the tensile properties of MMC deposits can be significantly improved.

Table 6.2 Tensile properties (strength and elongation) of as-fabricated cold sprayed MMC deposits

Powders	Weight ratio	Preparation method	Tensile strength (MPa)	Elongation (%)	References
Al + B_4C	60:40	Ball milling	37 ± 1.2	0.3 ± 0.03	[6, 41]
Al + B_4C	60:40	Ball milling	38 ± 3	0	[40]
Maraging steel 300 + WC	85:15	Mechanical blending	147.1 308	0.6 0.96	[44]
AA2024 + Al_2O_3	80:20	Mechanical blending	211	4.3	[42]
Al + Cladded Diamond	50:50	Mechanical blending	185.4	–	[22]
Al + Diamond	50:50	Mechanical blending	190.8	–	[22]

References

1. Yin, S., Cavaliere, P., Aldwell, B., Jenkins, R., Liao, H., Li, W., Lupoi, R.: Cold spray additive manufacturing and repair: fundamentals and applications. Addit. Manuf. **21**, 628–650 (2018). https://doi.org/10.1016/j.addma.2018.04.017
2. Li, W., Cao, C., Yin, S.: Solid-state cold spraying of Ti and its alloys: a literature review. Prog. Mater. Sci. **110**, 100633 (2020). https://doi.org/10.1016/j.pmatsci.2019.100633
3. Wang, Q., Spencer, K., Birbilis, N., Zhang, M.X.: The influence of ceramic particles on bond strength of cold spray composite coatings on AZ91 alloy substrate. Surf. Coat. Technol. **205**, 50–56 (2010). https://doi.org/10.1016/j.surfcoat.2010.06.008
4. Yu, M., Suo, X.K., Li, W.Y., Wang, Y.Y., Liao, H.L.: Microstructure, mechanical property and wear performance of cold sprayed Al5056/SiCp composite coatings: effect of reinforcement content. Appl. Surf. Sci. **289**, 188–196 (2014). https://doi.org/10.1016/j.apsusc.2013.10.132
5. Irissou, E., Legoux, J.G., Arsenault, B., Moreau, C.: Investigation of Al-Al_2O_3 cold spray coating formation and properties. J. Therm. Spray Technol. **16**, 661–668 (2007). https://doi.org/10.1007/s11666-007-9086-8
6. Tariq, N.H., Gyansah, L., Qiu, X., Du, H., Wang, J.Q., Feng, B., Yan, D.S., Xiong, T.Y.: Thermo-mechanical post-treatment: a strategic approach to improve microstructure and mechanical properties of cold spray additively manufactured composites. Mater. Des. **156**, 287–299 (2018). https://doi.org/10.1016/j.matdes.2018.06.062
7. Sansoucy, E., Marcoux, P., Ajdelsztajn, L., Jodoin, B.: Properties of SiC-reinforced aluminum alloy coatings produced by the cold gas dynamic spraying process. Surf. Coat. Technol. **202**, 3988–3996 (2008). https://doi.org/10.1016/j.surfcoat.2008.02.017
8. Peat, T., Galloway, A., Toumpis, A., McNutt, P., Iqbal, N.: The erosion performance of particle reinforced metal matrix composite coatings produced by co-deposition cold gas dynamic spraying. Appl. Surf. Sci. **396**, 1623–1634 (2016). https://doi.org/10.1016/j.apsusc.2016.10.155
9. Yin, S., Yan, X., Chen, C., Jenkins, R., Liu, M., Lupoi, R.: Hybrid additive manufacturing of Al-Ti6Al4V functionally graded materials with selective laser melting and cold spraying. J. Mater. Process. Technol. **255**, 650–655 (2018). https://doi.org/10.1016/j.jmatprotec.2018.01.015
10. Lupoi, R., Lupton, T., Jenkins, R., Robinson, A.J., O'Donnell, G.E.: Direct manufacturing of diamond composite coatings onto silicon wafers and heat transfer performance. CIRP Ann. **67**, 185–188 (2018). https://doi.org/10.1016/j.cirp.2018.04.092

11. Peat, T., Galloway, A., Toumpis, A., Steel, R., Zhu, W., Iqbal, N.: Enhanced erosion performance of cold spray co-deposited AISI316 MMCs modified by friction stir processing. Mater. Des. **120**, 22–35 (2017). https://doi.org/10.1016/j.matdes.2017.01.099
12. Luo, X.T., Yang, G.J., Li, C.J.: Multiple strengthening mechanisms of cold-sprayed cBNp/NiCrAl composite coating. Surf. Coat. Technol. **205**, 4808–4813 (2011). https://doi.org/10.1016/j.surfcoat.2011.04.065
13. Woo, D.J., Heer, F.C., Brewer, L.N., Hooper, J.P., Osswald, S.: Synthesis of nanodiamond-reinforced aluminum metal matrix composites using cold-spray deposition. Carbon N. Y. **86**, 15–25 (2015). https://doi.org/10.1016/j.carbon.2015.01.010
14. Ito, K., Ichikawa, Y.: Microstructure control of cold-sprayed pure iron coatings formed using mechanically milled powder. Surf. Coat. Technol. **357**, 129–139 (2018). https://doi.org/10.1016/j.surfcoat.2018.10.016
15. Pialago, E.J.T., Kwon, O.K., Kim, M.S., Park, C.W.: Ternary Cu-CNT-AlN composite coatings consolidated by cold spray deposition of mechanically alloyed powders. J. Alloys Compd. **650**, 199–209 (2015). https://doi.org/10.1016/j.jallcom.2015.08.007
16. Feng, C., Guipont, V., Jeandin, M., Amsellem, O., Pauchet, F., Saenger, R., Bucher, S., Iacob, C.: B4C/Ni Composite coatings prepared by cold spray of blended or CVD-coated powders. J. Therm. Spray Technol. **21**, 561–570 (2012). https://doi.org/10.1007/s11666-012-9774-x
17. Chen, C., Xie, X., Xie, Y., Planche, M.P., Deng, S., Ji, G., Aubry, E., Ren, Z., Liao, H.: Cold spraying of thermally softened Ni-coated FeSiAl composite powder: Microstructure characterization, tribological performance and magnetic property. Mater. Des. **160**, 270–283 (2018). https://doi.org/10.1016/j.matdes.2018.09.025
18. Huang, C., Li, W., Xie, Y., Planche, M.P., Liao, H., Montavon, G.: Effect of substrate type on deposition behavior and wear performance of Ni-coated graphite/Al composite coatings deposited by cold spraying. J. Mater. Sci. Technol. **33**, 338–346 (2017). https://doi.org/10.1016/j.jmst.2016.11.016
19. Yin, S., Xie, Y., Cizek, J., Ekoi, E.J., Hussain, T., Dowling, D.P., Lupoi, R.: Advanced diamond-reinforced metal matrix composites via cold spray: properties and deposition mechanism. Compos. Part B Eng. **113**, 44–54 (2017). https://doi.org/10.1016/j.compositesb.2017.01.009
20. Na, H., Bae, G., Shin, S., Kumar, S., Kim, H., Lee, C.: Advanced deposition characteristics of kinetic sprayed bronze/diamond composite by tailoring feedstock properties. Compos. Sci. Technol. **69**, 463–468 (2009). https://doi.org/10.1016/j.compscitech.2008.11.015
21. Johnson, T.G.: Metal coated particles and defense applications. In: Metal Coated Particles and Defense Applications. Cold Spray Action Team, Worcester (2012)
22. Yin, S., Cizek, J., Chen, C., Jenkins, R., O'Donnell, G., Lupoi, R.: Metallurgical bonding between metal matrix and core-shelled reinforcements in cold sprayed composite coating. Scr. Mater. **177**, 49–53 (2020). https://doi.org/10.1016/j.scriptamat.2019.09.023
23. Yin, S., Ekoi, E.J., Lupton, T.L., Dowling, D.P., Lupoi, R.: Cold spraying of WC-Co-Ni coatings using porous WC-17Co powders: formation mechanism, microstructure characterization and tribological performance. Mater. Des. **126**, 305–313 (2017). https://doi.org/10.1016/j.matdes.2017.04.040
24. Couto, M., Dosta, S., Guilemany, J.M.: Comparison of the mechanical and electrochemical properties of WC-17 and 12Co coatings onto Al7075-T6 obtained by high velocity oxy-fuel and cold gas spraying. Surf. Coat. Technol., pp. 180–189 (in press, 2014). https://doi.org/10.1016/j.surfcoat.2014.04.034
25. Li, W., Assadi, H., Gaertner, F., Yin, S.: A review of advanced composite and nanostructured coatings by solid-state cold spraying process. Crit. Rev. Solid State Mater. Sci. **44**, 109–156 (2019). https://doi.org/10.1080/10408436.2017.1410778
26. Sova, A., Kosarev, V.F., Papyrin, A., Smurov, I.: Effect of ceramic particle velocity on cold spray deposition of metal-ceramic coatings. J. Therm. Spray Technol. **20**, 285–291 (2011). https://doi.org/10.1007/s11666-010-9571-3
27. Grigoriev, S., Okunkova, A., Sova, A., Bertrand, P., Smurov, I.: Cold spraying: from process fundamentals towards advanced applications. Surf. Coat. Technol. **268**, 77–84 (2015). https://doi.org/10.1016/j.surfcoat.2014.09.060

28. Shkodkin, A., Kashirin, A., Klyuev, O., Buzdygar, T.: Metal particle deposition stimulation by surface abrasive treatment in gas dynamic spraying. J. Therm. Spray Technol. **15**, 382–386 (2006). https://doi.org/10.1361/105996306X124383
29. Yu, M., Li, W.Y., Suo, X.K., Liao, H.L.: Effects of gas temperature and ceramic particle content on microstructure and microhardness of cold sprayed SiCp/Al 5056 composite coatings. Surf. Coat. Technol. **220**, 102–106 (2013). https://doi.org/10.1016/j.surfcoat.2012.05.050
30. Yin, S., Hassani, M., Xie, Q., Lupoi, R.: Unravelling the deposition mechanism of brittle particles in metal matrix composites fabricated via cold spray additive manufacturing. Scr. Mater. **194**, 5 (2021). https://doi.org/10.1016/j.scriptamat.2020.10.055
31. Spencer, K., Fabijanic, D.M., Zhang, M.X.: The use of Al-Al_2O_3 cold spray coatings to improve the surface properties of magnesium alloys. Surf. Coat. Technol. **204**, 336–344 (2009). https://doi.org/10.1016/j.surfcoat.2009.07.032
32. Wu, J., Fang, H., Yoon, S., Kim, H., Lee, C.: The rebound phenomenon in kinetic spraying deposition. Scr. Mater. **54**, 665–669 (2006). https://doi.org/10.1016/j.scriptamat.2005.10.028
33. Assadi, H., Kreye, H., Gärtner, F., Klassen, T.: Cold spraying—a materials perspective. Acta Mater. **116**, 382–407 (2016). https://doi.org/10.1016/j.actamat.2016.06.034
34. Ichikawa, Y., Tokoro, R., Tanno, M., Ogawa, K.: Elucidation of cold-spray deposition mechanism by auger electron spectroscopic evaluation of bonding interface oxide fi lm. Acta Mater. **164**, 39–49 (2019). https://doi.org/10.1016/j.actamat.2018.09.041
35. Assadi, H., Gärtner, F., Stoltenhoff, T., Kreye, H.: Bonding mechanism in cold gas spraying. Acta Mater. **51**, 4379–4394 (2003). https://doi.org/10.1016/S1359-6454(03)00274-X
36. Guetta, S., Berger, M.H., Borit, F., Guipont, V., Jeandin, M., Boustie, M., Ichikawa, Y., Sakaguchi, K., Ogawa, K.: Influence of particle velocity on adhesion of cold-sprayed splats. J. Therm. Spray Technol. **18**, 331–342 (2009). https://doi.org/10.1007/s11666-009-9327-0
37. King, P.C., Bae, G., Zahiri, S.H., Jahedi, M., Lee, C.: An experimental and finite element study of cold spray copper impact onto two aluminum substrates. J. Therm. Spray Technol. **19**, 620–634 (2010). https://doi.org/10.1007/s11666-009-9454-7
38. Luo, X.T., Yang, E.J., Shang, F.L., Yang, G.J., Li, C.X., Li, C.J.: Microstructure, mechanical properties, and two-body abrasive wear behavior of cold-sprayed 2 vol.% Cubic BN-NiCrAl nanocomposite coating. J. Therm. Spray Technol. **23**, 1181–1190 (2014). https://doi.org/10.1007/s11666-014-0092-3
39. Shockley, J.M., Descartes, S., Vo, P., Irissou, E., Chromik, R.R.: The influence of Al2O3 particle morphology on the coating formation and dry sliding wear behavior of cold sprayed Al–Al_2O_3 composites. Surf. Coat. Technol. **270**, 324–333 (2015). https://doi.org/10.1016/j.surfcoat.2015.01.057
40. Meydanoglu, O., Jodoin, B., Kayali, E.S.: Microstructure, mechanical properties and corrosion performance of 7075 Al matrix ceramic particle reinforced composite coatings produced by the cold gas dynamic spraying process. Surf. Coat. Technol. **235**, 108–116 (2013). https://doi.org/10.1016/j.surfcoat.2013.07.020
41. Tariq, N.H., Gyansah, L., Wang, J.Q., Qiu, X., Feng, B., Siddique, M.T., Xiong, T.Y.: Cold spray additive manufacturing: a viable strategy to fabricate thick B4C/Al composite coatings for neutron shielding applications. Surf. Coat. Technol. **339**, 224–236 (2018). https://doi.org/10.1016/j.surfcoat.2018.02.007
42. Tariq, N. ul H., Gyansah, L., Qiu, X., Jia, C., Awais, H. Bin, Zheng, C., Du, H., Wang, J., Xiong, T.: Achieving strength-ductility synergy in cold spray additively manufactured Al/B 4 C composites through a hybrid post-deposition treatment. J. Mater. Sci. Technol. **35**, 1053–1063 (2019). https://doi.org/10.1016/j.jmst.2018.12.022
43. Yang, K., Li, W., Huang, C., Yang, X., Xu, Y.: Optimization of cold-sprayed AA2024/Al_2O_3 metal matrix composites via friction stir processing: effect of rotation speeds. J. Mater. Sci. Technol. **34**, 2167–2177 (2018). https://doi.org/10.1016/j.jmst.2018.03.016
44. Chen, C., Xie, Y., Yan, X., Huang, R., Kuang, M., Ma, W., Zhao, R., Wang, J., Liu, M., Ren, Z., Liao, H.: Cold sprayed WC reinforced maraging steel 300 composites: microstructure characterization and mechanical properties. J. Alloys Compd. **785**, 499–511 (2019). https://doi.org/10.1016/j.jallcom.2019.01.135

Chapter 7
Cold Sprayed Nanostructured Metallic Deposits

Abstract It is known that nanostructured materials generally have improved properties as compared with conventional materials. Therefore, it is of importance for CSAM to gain the capability to produce deposits with nanostructure. However, nanoparticles, due to their low weights, are difficult to deposit on the substrate via CSAM. They are easy to be picked up by the driving gas and thereby suffer from dramatic deceleration when passing through the compressed bow-shock in front of the substrate. Consequently, the impact velocity of nanoparticles is very low so that deposit is hard to form on the substrate. Fortunately, although nanoparticles cannot be deposited directly by CSAM, cold sprayed metal deposits can still exhibit nanostructure. Firstly, the starting feedstock for CSAM can be nanocrystalline powders; in this case, the deposit retains the nanostructure of the starting powders. Secondly, using nanomaterials to reinforce MMC deposits can also enable the deposit to present nanostructure. This chapter aims to provide an overview of the metal deposits with nanostructure produced by CSAM, particularly focusing on the deposit microstructure and properties.

7.1 Nanocrystalline Metal Deposits via CSAM

Nanocrystalline metals are polycrystalline metals with a crystallite size of a few nanometers (normally smaller than 100 nm). They generally exhibit increased strength and hardness, improved toughness, reduced elastic modulus and ductility, enhanced diffusivity, higher specific heat, enhanced thermal expansion coefficient, and superior soft magnetic properties in comparison with conventional polycrystalline metals [1]. CSAM, due to its low working temperature, has been found to be a robust tool to produce nanocrystalline metal deposits because nanocrystalline structure can be well retained in the deposit after deposition. As an evidence, Fig. 7.1 shows a TEM imaging of nanocrystalline Al 2018 deposit produced via CSAM, where nanocrystalline grains can be clearly observed in the deposit [2].

Preparation of nanocrystalline powders is one of the most important steps in the fabrication process. Mechanical ball milling is regarded as a simple, effective, and efficient method to produce nanocrystalline powders with a grain size as small as

S. Yin and R. Lupoi, *Cold Spray Additive Manufacturing*, Springer Tracts in Additive Manufacturing, https://doi.org/10.1007/978-3-030-73367-4_7

Fig. 7.1 TEM imaging of cold sprayed nanocrystalline Al 2018 deposit [2]

20 nm or below [3]. So far, various nanocrystalline powders have been produced through mechanical ball milling for CSAM [2, 4–13]. Table 7.1 summarizes the nanocrystalline powders produced for CSAM y and their milling conditions. As an example, Fig. 7.2 shows the EBSD IPF images of as-received Fe powder and ball-milled Fe powder. As can be seen, the grain size of the as-received Fe powder (Fig. 7.2a) is significantly refined into nano-size (Fig. 7.2b, c) after ball milling for 20 h. Ball milling also alter the shape of the powder from spherical to irregular [14].

Table 7.1 Nanocrystalline powders for CSAM and the milling conditions [2, 4-14]

Materials	Milling time, h	Speed, rpm	Ball diameter, mm	BPR	Control agent
Al alloy 2009	10	200	6.4	20:1	Liquid nitrogen
Al alloy 2618	8	180	6.4	32:1	Liquid nitrogen
Al alloy 5083	8	180	6.4	32:1	Liquid nitrogen
Al alloy 7075	N/A	180	11.6	32:1	Stearic acid
Al–Mg alloy	8	180	6.4	32:1	Liquid nitrogen
Ni	15	180	6.4	30:1	Liquid nitrogen
Ni–Ti alloy	48, 1/2 h rest per h	400	20	13:4	Alcohol
Ni-20Cr alloy	20, 1/3 h rest per h	300	N/A	10:2	Toluene
Cu	12	200	6.4	30:1	Liquid nitrogen
Fe	5, 10, 20	1400	8	73:50	N/A

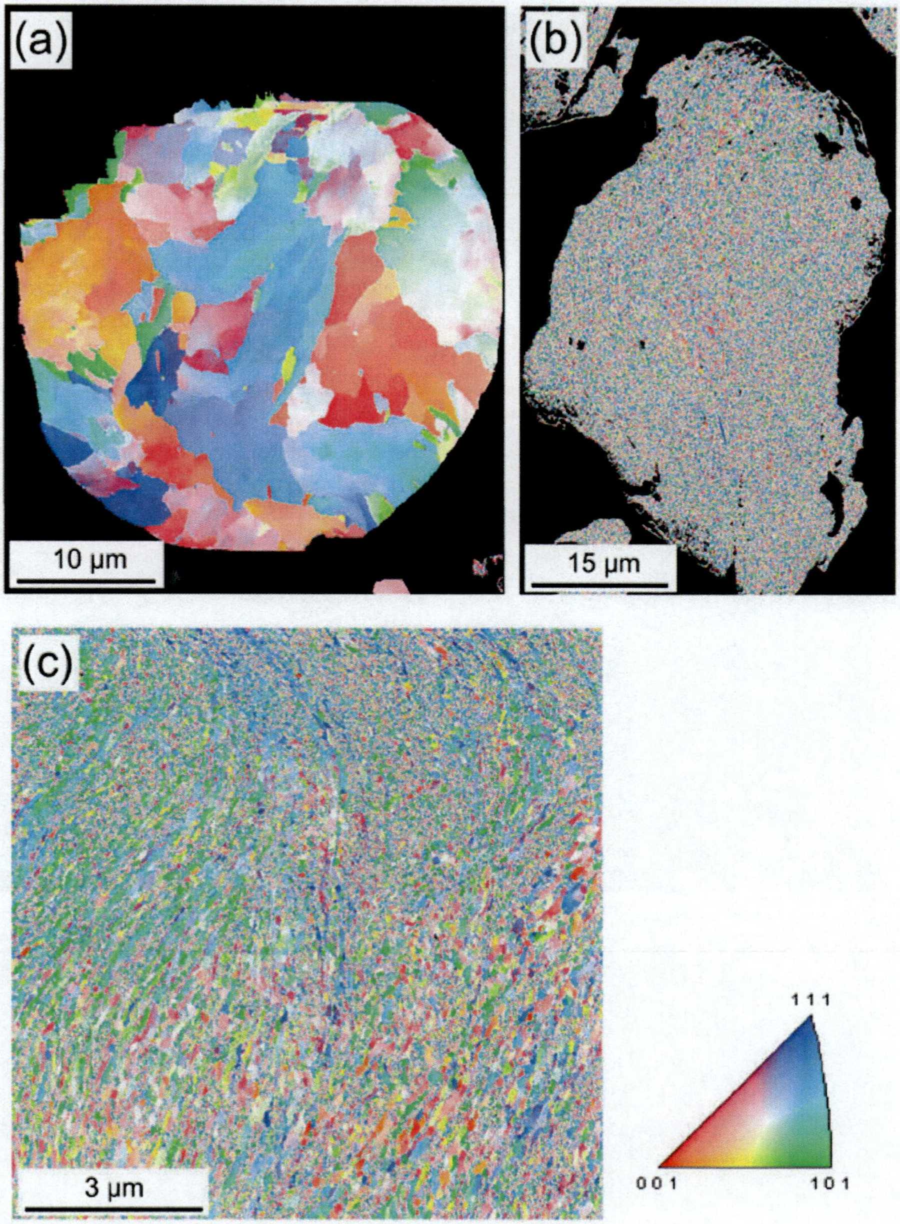

Fig. 7.2 EBSD IPF images of **a** as-received Fe powder, **b** ball-milled Fe powder in low-magnification, and **c** ball-milled Fe powder in high-magnification [14]

Due to the increase of grain boundaries, the nanocrystalline powder hardness was normally much higher than conventional counterpart, which makes the nanocrystalline deposit to be harder but more porous as compared to the conventional deposit [2, 11, 13]. For the same reason, the work hardening effect in the nanocrystalline deposits was not as prominent as in the conventional deposits [2, 7]. Figure 7.3 shows a comparison of cross-sectional microstructure between conventional and nanocrystalline Al deposits produced under the same working parameters [2]. It is seen that nanocrystalline deposit had higher porosity than conventional deposit due to the lack of sufficient plastic deformation. However, the deposit hardness showed an opposite trend; nanocrystalline deposit had a hardness of 4.41 GPa which was higher than the hardness of conventional deposit (3.75 GPa).

In terms of the mechanical properties, cold sprayed nanocrystalline deposits have shown better wear-resistance performance than conventional deposits as a result of higher hardness [11]. However, due to the simultaneous higher porosity, fatigue strength was found to not improve significantly [13]. So far, the property investigation

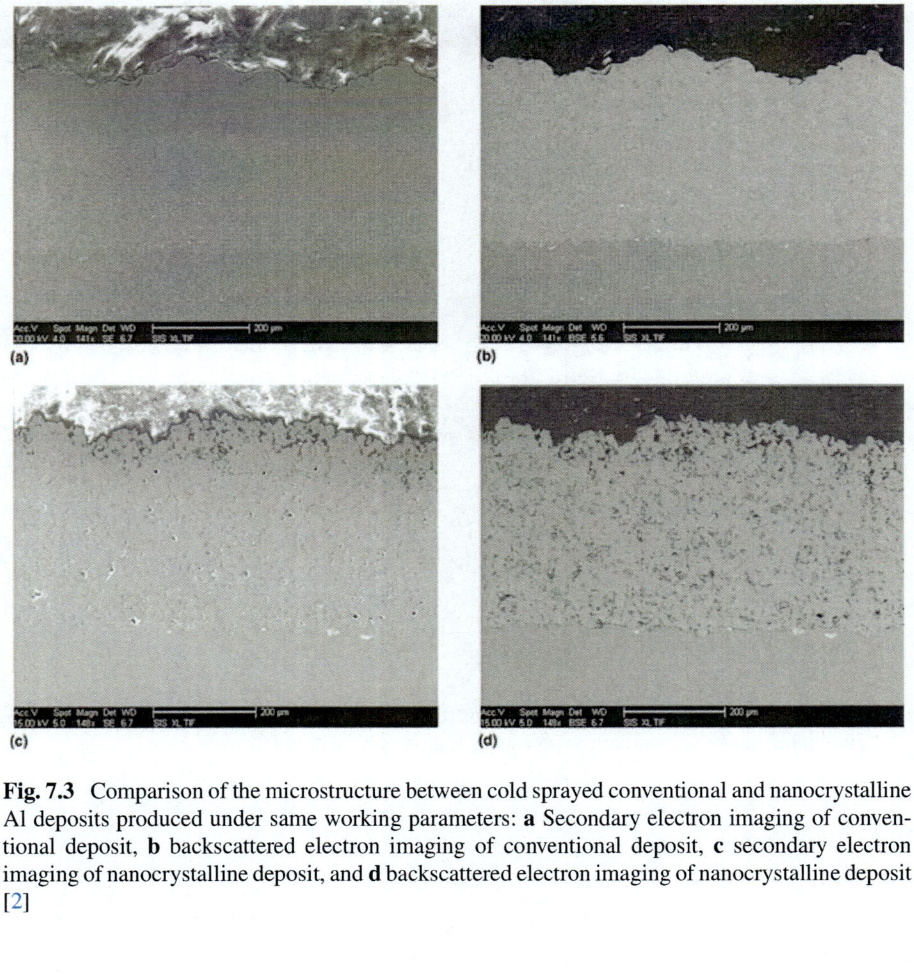

Fig. 7.3 Comparison of the microstructure between cold sprayed conventional and nanocrystalline Al deposits produced under same working parameters: **a** Secondary electron imaging of conventional deposit, **b** backscattered electron imaging of conventional deposit, **c** secondary electron imaging of nanocrystalline deposit, and **d** backscattered electron imaging of nanocrystalline deposit [2]

of cold sprayed nanocrystalline deposits are still very limited, e.g., investigations on deposit cohesion strength are still lacking. Considering the unique advantages of hard nanocrystalline deposits, more mechanical property tests such as deposit tensile stress and elongation are encouraged in the future work.

7.2 1D-Material-Reinforced MMC Deposits via CSAM

Carbon nanotube (CNT) is an allotrope of carbon with a cylindrical nanostructure. The diameter of CNT can be as small as 1 nm but the length can be up to several centimeters, thus CNT is also recognized as 1D nanomaterial. As a member of carbon family, CNT has extraordinary thermal conductivity, electrical and mechanical properties. These novel properties make CNT potentially valuable and useful in a wide variety of applications in nanotechnology, electronics, optics, thermal engineering and other fields of material science [15]. Due to the 1D nanostructure, CNT cannot exist in the form of bulk state. Thereby, it is widely employed as reinforcements for improving the properties of pure metals. Currently, CNT-reinforced MMCs are mainly produced by powder metallurgy [16, 17] and thermal spray technologies [18, 19]. These processes generally require high temperature to melt the binder phase, resulting in damage and phase transformation of CNT during fabrication [20, 21]. CSAM, due to its low working temperature, has been applied to produce CNT-reinforced MMC deposits in recent years.

A number of investigations have proved that dense and thick CNT-reinforced MMC deposits can be fabricated via CSAM [22–27]. Among all these works, mechanical ball milling as a robust technology was prevailingly used to prepare CNT-reinforced MMC powders. Ball milling is a process where mixed powders is placed in a chamber and subjected to high-energy collision from balls to induce mechanical alloying. Figure 7.4 shows the cross-sectional view and EDX mapping of a ball-milled CNT-Cu MMC powder. As can be seen, CNT was successfully incorporated into the MMC powder and exhibited a homogenous distribution [22]. Following the preparation of the MMC powders, various CNT-reinforced MMC deposits were fabricated as listed in Table 7.2 [22–25, 28–31]. As an example, Fig. 7.5 shows the microstructure characterizations of cold sprayed CNT-AlSi deposit. Clearly, the deposit is very thick and dense with little to no defects as shown in Fig. 7.5a. The TEM analysis shown in Fig. 7.5b–d shows that CNT was successfully deposited in the cold sprayed Al-based MMC deposit [27].

Although mechanical ball milling is promising for producing CNT-reinforced MMC powders, it also brings negative aspects. CNT reinforcements suffered from damage during the milling process due to the plastic deformation of binder phase [28]. In addition, the high-velocity impact happing during the deposit deposition process also led to the fracture of CNT [25, 28]. Figure 7.6 shows the TEM imaging of damaged CNT in the cold sprayed CNT-Cu deposits [25]. Two different damaging features can be noticed, which are impact-induced and shear-induced damages, respectively. The impact-induced damage was present in the form of a systematic

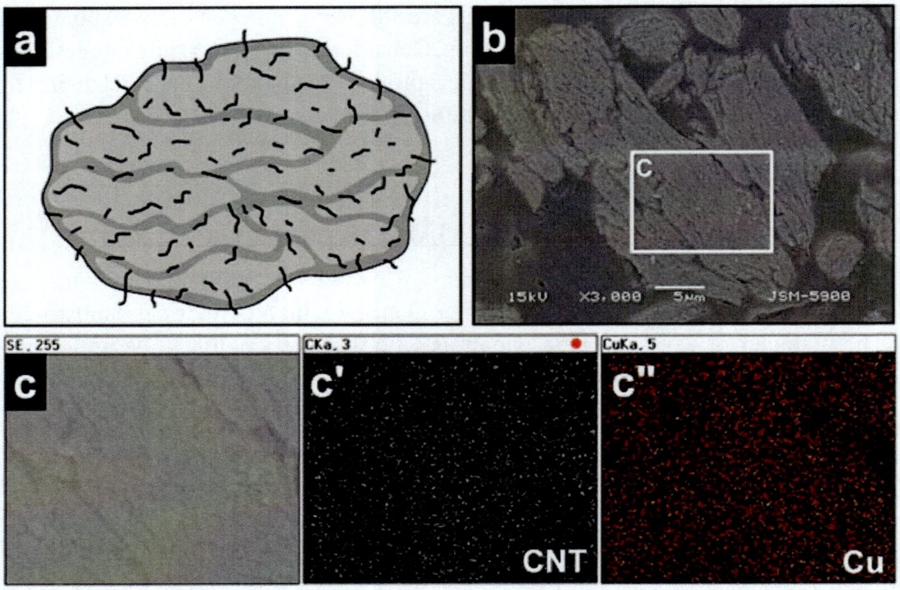

Fig. 7.4 Cross-sectional view and EDX mapping of a ball-milled CNT-Cu MMC powder [22]

Table 7.2 CNT-reinforced MMC deposits produced by CSAM and their thermal properties [22–25, 28–31]

Composites	Contents of CNT	Properties	References
CNT-Al	0.5–1.0% by weight	N/A	[25]
CNT-Cu	3% by volume	Higher thermal diffusivity	[28]
CNT-Cu	5–15% by volume	Better heat transfer performance	[22, 29]
CNT-Cu-SiC	4–4.5% by volume	Better heat transfer performance	[24, 30, 31]
CNT-Cu-AlN	4–4.5% by volume	Better heat transfer performance	[23, 30, 31]
CNT-Cu-BN	4% by volume	Better heat transfer performance	[30, 31]
CNT-AlSi	1.5% by weight	N/A	[27]
CNT-Al	1% by weight	Higher hardness and better wear-/corrosion-resistance	[26]
CNT-Al	1.5% by weight	Higher hardness	[32]

fracture of the concentric tubes which progresses inward until the innermost tube has broken, while the shear-induced damage was featured uneven or asymmetric with respect to tube axis [25]. Currently, prevention of damage of CNT during ball milling process is still a challenging work, which may be a research focus in the future work.

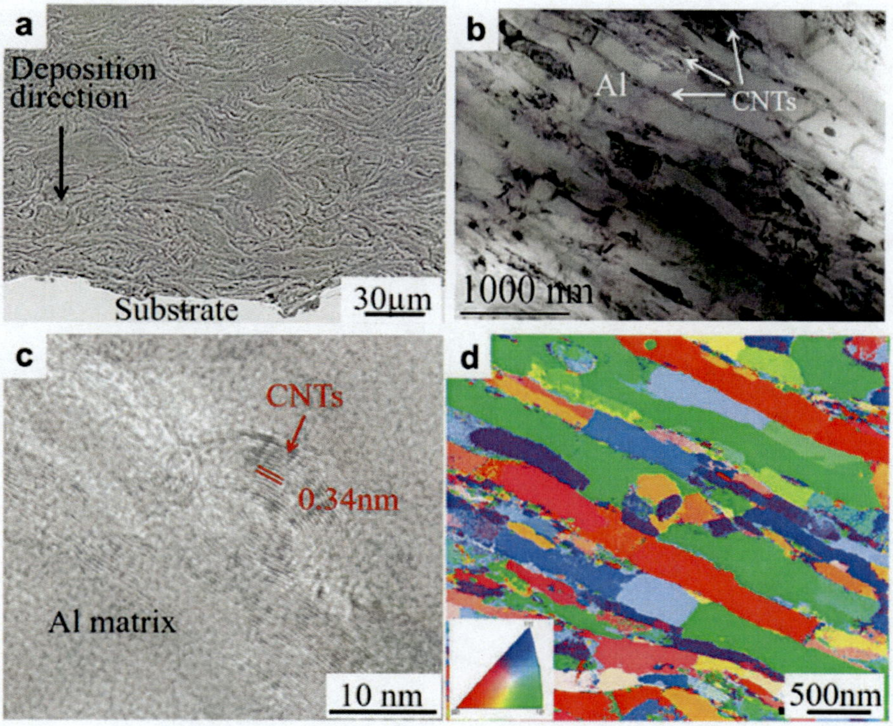

Fig. 7.5 Microstructure characterizations of cold sprayed CNT-AlSi deposit: **a** Cross-sectional microstructure of the deposit, **b** TEM image showing the nanolaminated structure in the CNT/AlSi composite; **c** HRTEM image highlighting the CNTs in the Al matrix; **d** TEM/EBSD orientation map [27]

In addition to mechanical ball milling, spray drying technology has also been used to disperse CNT within agglomerated metal powders. Spray drying is a process that produces powders from a liquid solution by rapid drying with hot gases. With this method, CNT was mainly embedded on the surface of agglomerated metal powders but hard to be homogenously incorporated inside the individual powder. Figure 7.7 shows the schematic of the deposit fabrication procedure using spray drying MMC powders. As can be seen, a consequence of using spray drying powders is the inhomogeneous distribution of CNT in the cold sprayed deposits as shown in Fig. 7.8 [25].

As for the properties of cold sprayed CNT-reinforced MMC deposits, it has been revealed that mechanical properties improved with additional CNT reinforcements. In the CNT-Al MMC deposits, local hardness was found to be higher at CNT-rich zone due to the higher stiffness of CNT [25]. Similar results were also concluded in the CNT-Cu MMC deposits. CNT reinforcements were also found to lead to the improvement of thermal properties in terms of both heat transfer performance and thermal diffusivity [28, 29]. Table 7.2 lists the existing CNT-reinforced MMC deposits produced by CSAM and their improved thermal properties [22–25, 28–31].

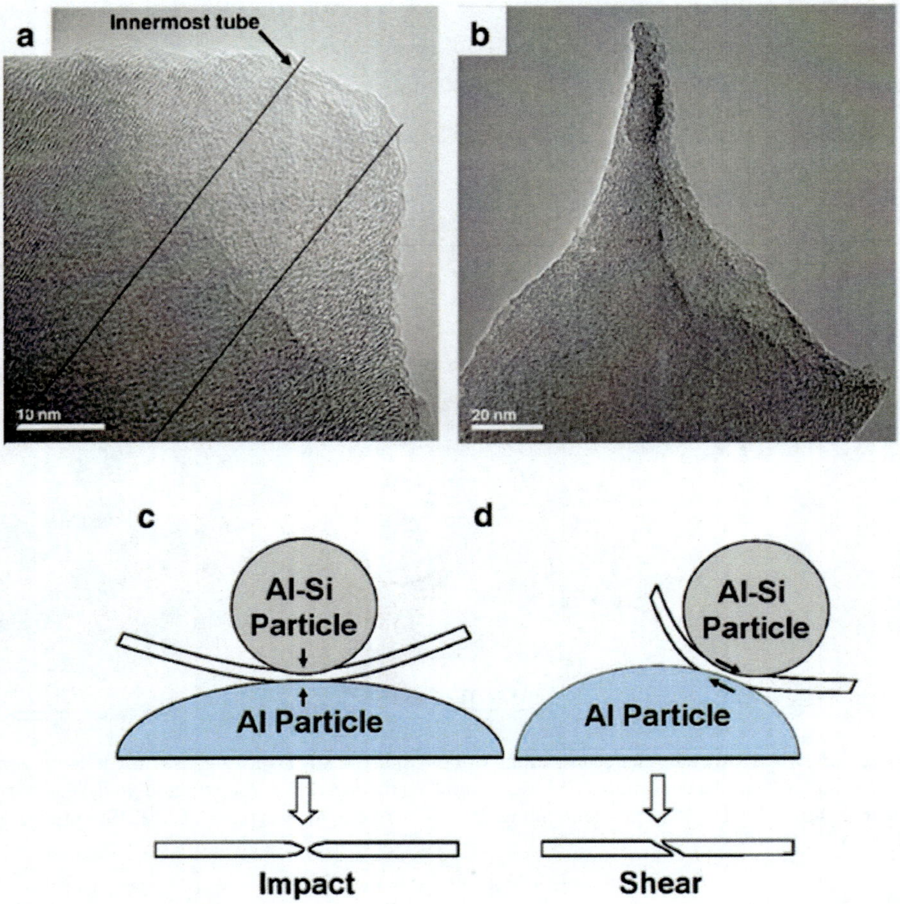

Fig. 7.6 TEM imaging of damaged CNT in the cold sprayed CNT-Cu deposits and their damaging mechanism. **a**, **c** impact-induced damage, and **b**, **d** shear-induced damage [25]

The existing works clearly demonstrate that CSAM is a promising technology to produce CNT-reinforced MMC deposits.

7.3 2D-Material-Reinforced MMC Deposits via CSAM

7.3.1 Graphene-Reinforced MMC Deposits

Graphene, as the single layer of sp^2 bonded carbon atoms, has extraordinary mechanical, thermal and electrical properties, attracting great attentions from both scientific and industrial communities. It is normally applied as reinforcements of MMCs to

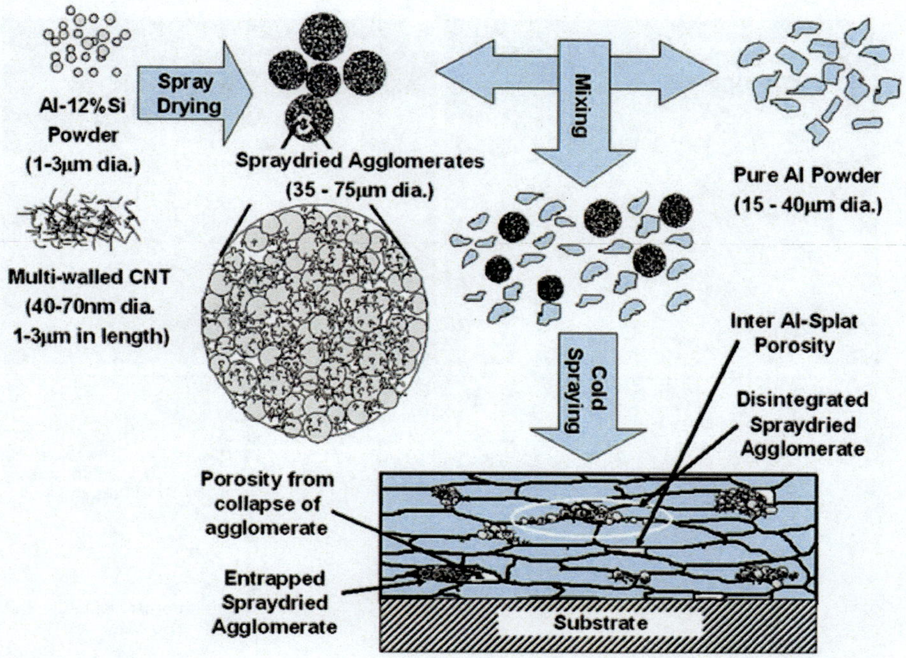

Fig. 7.7 Schematic of the deposit fabrication procedure using spray drying MMC powders [25]

improve the matrix material properties. The graphene-reinforced MMCs have exhibited superior properties over the pure metals [33]. Currently, the most common ways for fabricating graphene-reinforced MMCs are a number of powder metallurgy techniques, e.g., spark plasma sintering, laser sintering, hot pressing [34–39]. The existing works showed great capability of graphene-reinforced MMCs to improve the materials properties, e.g. strength [34–36], Young's modulus [36, 40], hardness [36, 40], wear-resistance [36–38], and electrical conductivity [39].

Graphene-reinforced MMCs were successfully produced via CSAM in very recent years [41, 42]. As a key step, preparation of MMC powders is of great importance to the deposit deposition process and final deposit quality. So far, two manufacturing methods have been used for MMC powder fabrication. In Yin et al.'s work, mechanical ball milling was applied to incorporate graphene nanosheets into spherical Cu particles [41]. Because the energy for ball milling is not high in that work, graphene nanosheets were mainly embedded on the Cu particle surface rather than homogenously distributed inside the Cu particle. Alternatively, in the work of Dardona et al., the graphene-reinforced MMC powders were synthesized through electroless plating of Cu film on the surface of graphene nanosheets [42]. Electroless plating is a method that deposits metals onto a solid piece by chemical approach. The piece to be plated is immersed in a reducing agent where metal ions can be changed to metal when catalyzed by certain materials to form a deposit. Figure 7.9 shows the morphology of the graphene-reinforced Cu MMC powders produced through electorless plating

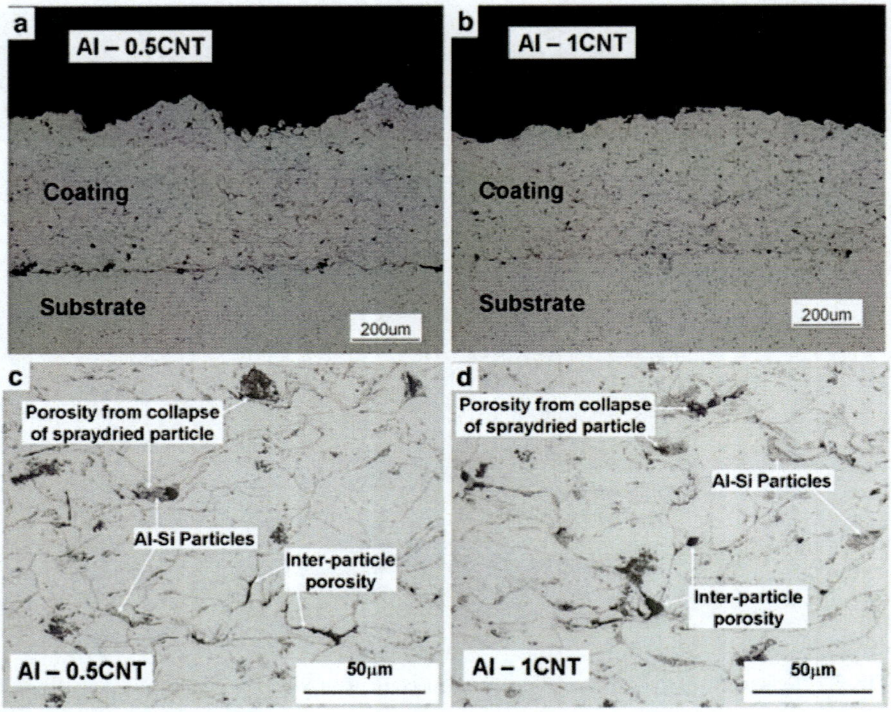

Fig. 7.8 SEM imaging of the cross-section of cold sprayed CNT-Al MMC deposits with different CNT contents. **a** CNT content of 0.5%, **b** CNT content of 1%, and **c**, **d** high-magnification view [25]

under different plating time. Powders produced in this way basically have a graphene core and an outside Cu film, which is opposite to the ball milled powders. Longer plating time would result in more Cu phase on the graphene surface as can be seen from Fig. 7.9.

As for the deposit microstructure, ball milled MMC powders resulted in dense and thick deposit with uniformly distributed graphene nanosheets as shown Fig. 7.10. However, the electroless plated MMC powders seemed to produce a low quality deposit with insufficient cohesion strength and inhomogeneous distribution of graphene nanosheets in the deposit as shown in Fig. 7.11 [41]. This may be due to the low fraction of Cu phase in the MMC powders, which significantly limits the effective metallic bonding between inter-particles [42]. In terms of the deposit properties, currently, only the wear-resistance and electrical conductivity have been tested. In Yin et al.'s work, the graphene-reinforced MMC deposits exhibited excellent wear-resistance performance, better than those produced by spark plasma sintering [41]. In the work of Dardona et al., the deposit demonstrated worse electrical conductivity than bulk Cu, which was probably due to the low Cu thickness on the graphene powder surface, poor inter-particle bonding and inhomogeneous distribution of graphene nanosheets in the deposit [42].

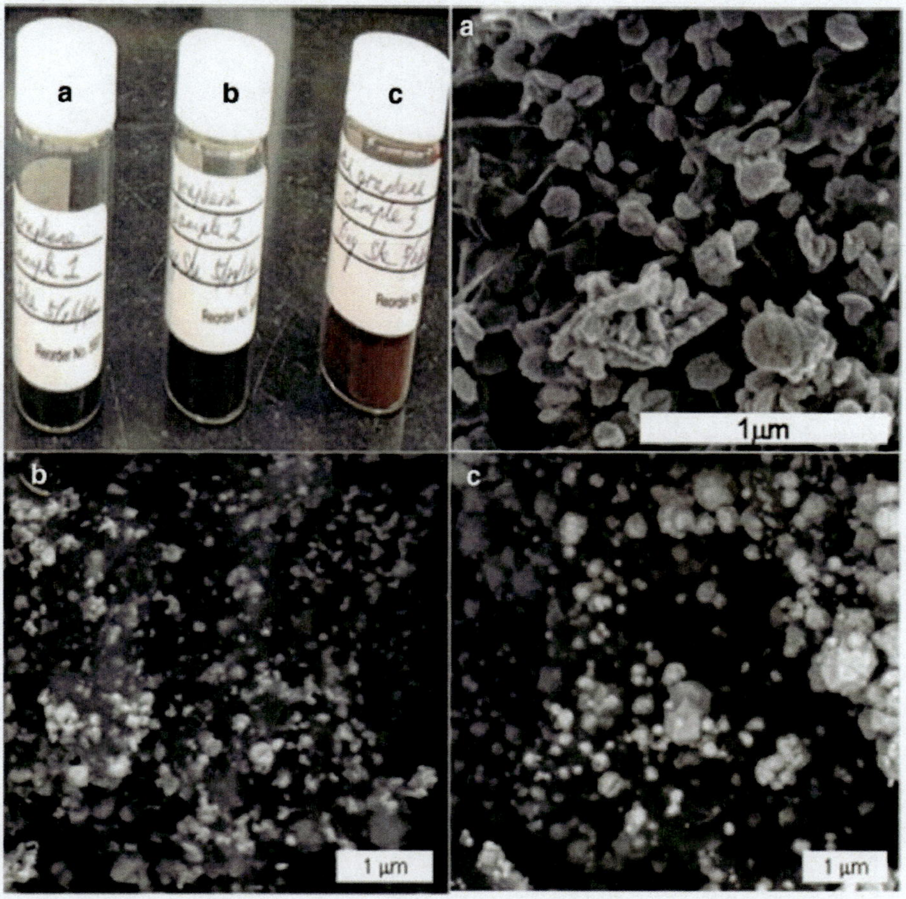

Fig. 7.9 Morphology of the graphene-Cu MMC powders produced through electorless plating under different plating time. **a** Digital photo of all samples, **b** plating for 1.5 h, **c** plating for 3 h, and **d** plating for 4.5 h [42]

7.3.2 WS₂-Reinforced MMC Deposits

Monolayer tungsten disulfide (WS_2) has great potential in the optical sector due to its direct band gap and high photoluminescence intensity [43]. It also possesses excellent solid lubrication properties due to the 2D-layered structure and easy interlayer sliding. In addition, the special structure of WS_2 allows it to be usable in high temperature, high pressure, high vacuum, high load and with radiation and corrosive media environments [44, 45]. Moreover, because WS_2 has an excellent adsorption capacity on the metal surface, it can be used as reinforcements in MMCs to improve the lubrication performance [46]. The WS_2-Al MMCs made by sparkling plasma sintering have shown better wear-resistance performance than graphene-Al MMCs

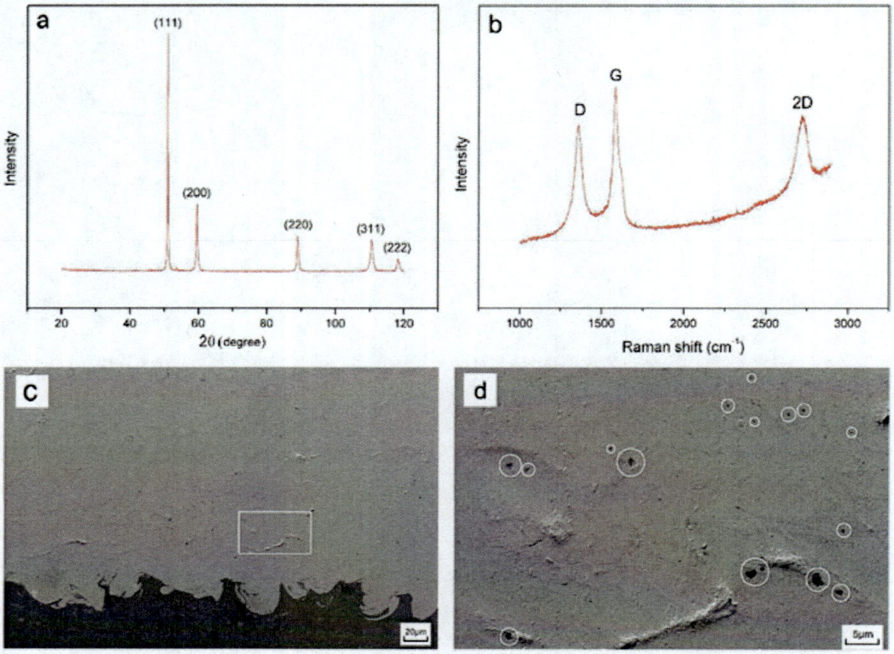

Fig. 7.10 Characterization of graphene-Cu MMC deposit produced using ball milled powders: **a** XRD spectrum, **b** Raman spectrum, **c** deposit cross-sectional view, and **d** a magnified view [41]

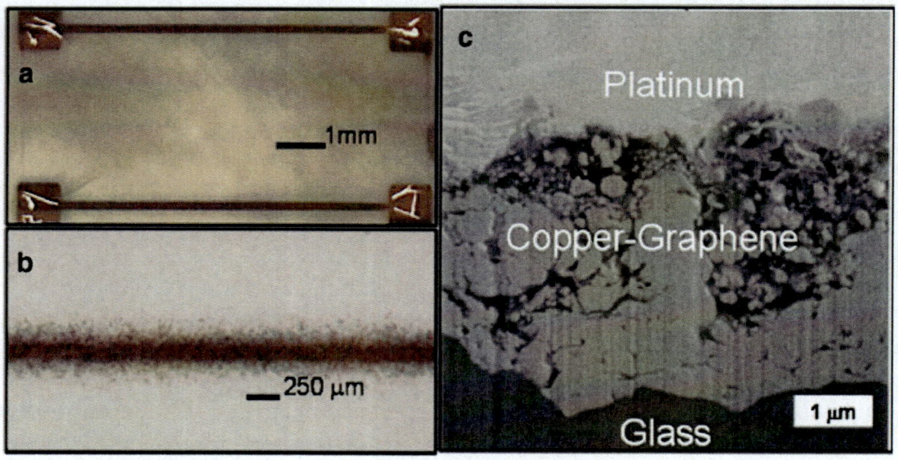

Fig. 7.11 Characterization of graphene-Cu MMC deposit produced using electroless plated powders: **a** Optical view of two tracks, **b** single-line track, and **c** deposit cross-sectional view [42]

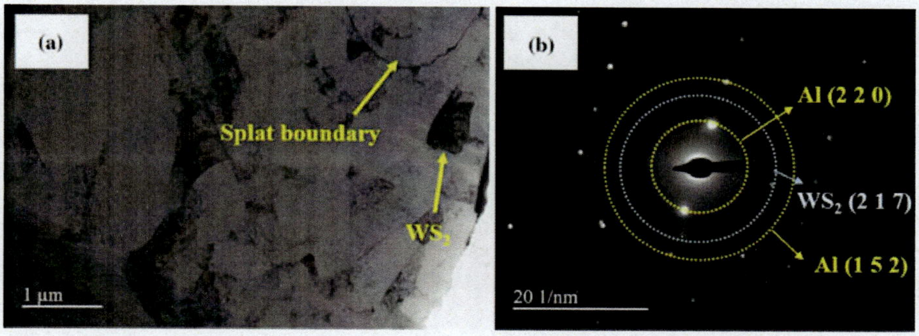

Fig. 7.12 TEM imaging of WS_2-Al MMC deposit: **a** Bright field imaging with multiple splats, sub-grains and **b** SAED pattern [47]

under ball-on-disk wear tests, which clearly demonstrates the superiority of WS_2 than other lubricants [45].

WS_2-Al MMC deposits were produced via CSAM for wear resistance in a recent work [47]. Ball milling technology was used to produce the MMC powders. After low-energy ball milling, Al particles still remained spherical shape, and WS_2 was mostly attached on the surface of the Al particles. The surface morphology of the WS_2-Al MMC powders is quite similar with the graphene-Cu powders used in Yin et al.'s study [41]. The deposits were then produced using nitrogen under the pressure of 3.8 MPa and temperature of 400 °C. Figure 7.12 shows the TEM imaging of the deposit cross-section. As can be seen, WS_2 was successfully deposited with Al onto the carbon steel substrate and uniformly distributed within the deposit, which demonstrates the feasibility of CSAM to produce WS_2-reinforced MMC deposits. The wear test revealed that the WS_2-reinforced MMC deposit had an outstanding wear-resistance performance due to the presence of 2D layered WS_2 which aids in shearing of WS_2 layers and uniform tribofilm formation comprised of WS_2 and WO_3.

7.3.3 hBN-Reinforced MMC Deposits

Hexagonal boron nitride (hBN) nanosheets are 2D crystalline form of hBN, which have a thickness of one to few atomic layers. It is similar in geometry to graphene, but having completely different chemical, thermal and electronic properties. CSAM has been successfully used for producing hBN-reinforced MMC deposits, mainly hBN-Ni deposits [48–51]. Electroless plating was used to encapsulate hBN powders and to produce hBN-reinforced MMC powders. During the encapsulation process, the MMC particles tended to agglomerate and form large clusters. Therefore, as a comparison, following the electroless plating, ball milling were employed to de-agglomerate the clustered feedstock powders [49]. Both low- and high- energy ball milling methods were employed: low energy can eliminate voids inside the clusters and lead to higher density and uniform particle size; high energy resulted in

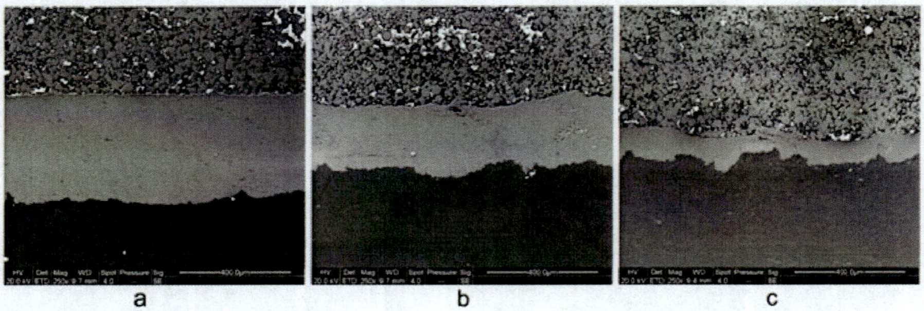

Fig. 7.13 SEM imaging of the cross-section of cold sprayed hBN-Ni MMC deposits produced using **a** Unaltered powders, **b** ball milled powders for 1 h, and **c** ball milled powders for 2 h [50]

breakup of agglomerations and destroyed the Ni encapsulant. Figure 7.13 shows the SEM imaging of cross-section of cold sprayed hBN-Ni MMC deposits using various powders. As can be seen, the unaltered hBN-Ni MMC powders resulted in the thickest and densest deposits due to the significant plastic deformation as compared with ball milled powders which experienced work hardening during the powder preparation process [50]. The tribological study demonstrates that the cold sprayed hBN-Ni MMC deposits had a very promising wear-resistance performance. They can reduce the friction coefficient by almost 50% and significantly increase the wear resistance as compared with pure Ni [51].

References

1. Meyers, M.A., Mishra, A., Benson, D.J.: Mechanical properties of nanocrystalline materials. Prog. Mater. Sci. **51**, 427–556 (2006). https://doi.org/10.1016/j.pmatsci.2005.08.003
2. L., Zúñiga, A., Jodoin, B., Lavernia, E.J.: Cold-spray processing of a nanocrystalline Al-Cu-Mg-Fe-Ni alloy with Sc. J. Therm. Spray Technol. 15, 184–190 (2006). https://doi.org/10.1361/105996306X107995
3. Gusev, A.I., Rempel, A.A.: Nanocrystalline Materials. Cambridge International Science Publishing (2004)
4. Ajdelsztajn, L., Jodoin, B., Kim, G.E., Schoenung, J.M.: Cold spray deposition of nanocrystalline aluminum alloys. Metall. Mater. Trans. A Phys. Metall. Mater. Sci. **36**, 657–666 (2005). https://doi.org/10.1007/s11661-005-0099-y
5. Kumar, M., Singh, H., Singh, N., Hong, S.M., Choi, I.S., Suh, J.Y., Chavan, N.M., Kumar, S., Joshi, S.V.: Development of nano-crystalline cold sprayed Ni-20Cr coatings for high temperature oxidation resistance. Surf. Coat. Technol. **266**, 122–133 (2015). https://doi.org/10.1016/j.surfcoat.2015.02.032
6. Ajdelsztajn, L., Jodoin, B., Schoenung, J.M.: Synthesis and mechanical properties of nanocrystalline Ni coatings produced by cold gas dynamic spraying. Surf. Coat. Technol. **201**, 1166–1172 (2006). https://doi.org/10.1016/j.surfcoat.2006.01.037
7. Richer, P., Jodoin, B., Ajdelsztajn, L., Lavernia, E.J.: Substrate roughness and thickness effects on cold spray nanocrystalline Al-Mg coatings. J. Therm. Spray Technol. **15**, 246–254 (2006). https://doi.org/10.1361/105996306X108174

8. Sudharshan Phani, P., Vishnukanthan, V., Sundararajan, G.: Effect of heat treatment on properties of cold sprayed nanocrystalline copper alumina coatings. Acta Mater. **55**, 4741–4751 (2007). https://doi.org/10.1016/j.actamat.2007.04.044
9. Zhang, Y.Y., Wu, X.K., Cui, H., Zhang, J.S.: Cold-spray processing of a high density nanocrystalline aluminum alloy 2009 coating using a mixture of as-atomized and as-cryomilled powders. J. Therm. Spray Technol. **20**, 1125–1132 (2011). https://doi.org/10.1007/s11666-011-9652-y
10. Tria, S., Elkedim, O., Hamzaoui, R., Guo, X., Bernard, F., Millot, N., Rapaud, O.: Deposition and characterization of cold sprayed nanocrystalline NiTi. Powder Technol. **210**, 181–188 (2011). https://doi.org/10.1016/j.powtec.2011.02.026
11. Liu, J., Zhou, X., Zheng, X., Cui, H., Zhang, J.: Tribological behavior of cold-sprayed nanocrystalline and conventional copper coatings. Appl. Surf. Sci. **258**, 7490–7496 (2012). https://doi.org/10.1016/j.apsusc.2012.04.070
12. Liu, J., Cui, H., Zhou, X., Wu, X., Zhang, J.: Nanocrystalline copper coatings produced by cold spraying. Met. Mater. Int. **18**, 121–128 (2012). https://doi.org/10.1007/s12540-012-0014-1
13. Ghelichi, R., Bagherifard, S., Mac Donald, D., Brochu, M., Jahed, H., Jodoin, B., Guagliano, M.: Fatigue strength of Al alloy cold sprayed with nanocrystalline powders. Int. J. Fatigue **65**, 51–57 (2014). https://doi.org/10.1016/j.ijfatigue.2013.09.001
14. Ito, K., Ichikawa, Y.: Microstructure control of cold-sprayed pure iron coatings formed using mechanically milled powder. Surf. Coat. Technol. **357**, 129–139 (2018). https://doi.org/10.1016/j.surfcoat.2018.10.016
15. Wang, X., Li, Q., Xie, J., Jin, Z., Wang, J., Li, Y., Jiang, K., Fan, S.: Fabrication of ultralong and electrically uniform single-walled carbon nanotubes on clean substrates. Nano Lett. **9**, 3137–3141 (2009). https://doi.org/10.1021/nl901260b
16. Song, J.L., Chen, W.G., Dong, L.L., Wang, J.J., Deng, N.: An electroless plating and planetary ball milling process for mechanical properties enhancement of bulk CNTs/Cu composites. J. Alloys Compd. **720**, 54–62 (2017). https://doi.org/10.1016/j.jallcom.2017.05.208
17. Cheng, B., Bao, R., Yi, J., Li, C., Tao, J., Liu, Y., Tan, S., You, X.: Interface optimization of CNT/Cu composite by forming TiC nanoprecipitation and low interface energy structure via spark plasma sintering. J. Alloys Compd. **722**, 852–858 (2017). https://doi.org/10.1016/j.jallcom.2017.06.186
18. Kaewsai, D., Watcharapasorn, A., Singjai, P., Wirojanupatump, S., Niranatlumpong, P., Jiansirisomboon, S.: Thermal sprayed stainless steel/carbon nanotube composite coatings. Surf. Coat. Technol. **205**, 2104–2112 (2010). https://doi.org/10.1016/j.surfcoat.2010.08.113
19. Moonngam, S., Tunjina, P., Deesom, D., Banjongprasert, C.: Fe-Cr/CNTs nanocomposite feedstock powders produced by chemical vapor deposition for thermal spray coatings. Surf. Coat. Technol. **306**, 323–327 (2016). https://doi.org/10.1016/j.surfcoat.2016.07.024
20. Curtin, W.A., Sheldon, B.W.: CNT-reinforced ceramics and metals. Mater. Today. **7**, 44–49 (2004). https://doi.org/10.1016/S1369-7021(04)00508-5
21. Dorri Moghadam, A., Omrani, E., Menezes, P.L., Rohatgi, P.K.: Mechanical and tribological properties of self-lubricating metal matrix nanocomposites reinforced by carbon nanotubes (CNTs) and grapheme—a review. Compos. Part B Eng. **77**, 402–420 (2015). https://doi.org/10.1016/j.compositesb.2015.03.014
22. Pialago, E.J.T., Park, C.W.: Cold spray deposition characteristics of mechanically alloyed Cu-CNT composite powders. Appl. Surf. Sci. **308**, 63–74 (2014). https://doi.org/10.1016/j.apsusc.2014.04.096
23. Pialago, E.J.T., Kwon, O.K., Kim, M.S., Park, C.W.: Ternary Cu-CNT-AlN composite coatings consolidated by cold spray deposition of mechanically alloyed powders. J. Alloys Compd. **650**, 199–209 (2015). https://doi.org/10.1016/j.jallcom.2015.08.007
24. Pialago, E.J.T., Kwon, O.K., Park, C.W.: Cold spray deposition of mechanically alloyed ternary Cu-CNT-SiC composite powders. Ceram. Int. **41**, 6764–6775 (2015). https://doi.org/10.1016/j.ceramint.2015.01.123
25. Bakshi, S.R., Singh, V., Balani, K., McCartney, D.G., Seal, S., Agarwal, A.: Carbon nanotube reinforced aluminum composite coating via cold spraying. Surf. Coat. Technol. **202**, 5162–5169 (2008). https://doi.org/10.1016/j.surfcoat.2008.05.042

26. Zhang, Y., Wang, Q., Chen, G., Ramachandran, C.S.: Mechanical, tribological and corrosion physiognomies of CNT-Al metal matrix composite (MMC) coatings deposited by cold gas dynamic spray (CGDS) process. Surf. Coat. Technol. **403**, 126380 (2020). https://doi.org/10.1016/j.surfcoat.2020.126380
27. Xie, X., Chen, C., Ji, G., Xu, R., Tan, Z., Xie, Y., Li, Z., Liao, H.: A novel approach for fabricating a CNT/AlSi composite with the self-aligned nacre-like architecture by cold spraying. Nano Mater. Sci. **1**, 137–141 (2019). https://doi.org/10.1016/j.nanoms.2019.04.002
28. Cho, S., Takagi, K., Kwon, H., Seo, D., Ogawa, K., Kikuchi, K., Kawasaki, A.: Multi-walled carbon nanotube-reinforced copper nanocomposite coating fabricated by low-pressure cold spray process. Surf. Coat. Technol. **206**, 3488–3494 (2012). https://doi.org/10.1016/j.surfcoat.2012.02.021
29. Pialago, E.J.T., Kwon, O.K., Park, C.W.: Nucleate boiling heat transfer of R134a on cold sprayed CNT-Cu composite coatings. Appl. Therm. Eng. **56**, 112–119 (2013). https://doi.org/10.1016/j.applthermaleng.2013.03.046
30. Pialago, E.J.T., Kwon, O.K., Jin, J.S., Park, C.W.: Nucleate pool boiling of R134a on cold sprayed Cu-CNT-SiC and Cu-CNT-AlN composite coatings. Appl. Therm. Eng. **103**, 684–694 (2016). https://doi.org/10.1016/j.applthermaleng.2016.04.022
31. Pialago, E.J., Ada, K.A., Park, C.W.: MNHMT2012–75261. 6–13 (2016)
32. Xie, X., Tan, Z., Chen, C., Xie, Y., Wu, H., Yan, X., Gao, S., Li, Z., Ji, G., Liao, H.: Synthesis of carbon nanotube reinforced Al matrix composite coatings via cold spray deposition. Surf. Coat. Technol. **405**, 126676 (2020). https://doi.org/10.1016/j.surfcoat.2020.126676
33. Lin, D., Richard Liu, C., Cheng, G.J.: Single-layer graphene oxide reinforced metal matrix composites by laser sintering: microstructure and mechanical property enhancement. Acta Mater. **80**, 183–193 (2014). https://doi.org/10.1016/j.actamat.2014.07.038
34. Chen, L.Y., Konishi, H., Fehrenbacher, A., Ma, C., Xu, J.Q., Choi, H., Xu, H.F., Pfefferkorn, F.E., Li, X.C.: Novel nanoprocessing route for bulk graphene nanoplatelets reinforced metal matrix nanocomposites. Scr. Mater. **67**, 29–32 (2012). https://doi.org/10.1016/j.scriptamat.2012.03.013
35. Chen, F., Ying, J., Wang, Y., Du, S., Liu, Z., Huang, Q.: Effects of graphene content on the microstructure and properties of copper matrix composites. Carbon N. Y. **96**, 836–842 (2016). https://doi.org/10.1016/j.carbon.2015.10.023
36. Zhai, W., Shi, X., Wang, M., Xu, Z., Yao, J., Song, S., Wang, Y.: Grain refinement: a mechanism for graphene nanoplatelets to reduce friction and wear of Ni3Al matrix self-lubricating composites. Wear **310**, 33–40 (2014). https://doi.org/10.1016/j.wear.2013.12.014
37. Xu, Z., Shi, X., Zhai, W., Yao, J., Song, S., Zhang, Q.: Preparation and tribological properties of TiAl matrix composites reinforced by multilayer graphene. Carbon N. Y. **67**, 168–177 (2014). https://doi.org/10.1016/j.carbon.2013.09.077
38. Fan, Y., Kang, L., Zhou, W., Jiang, W., Wang, L., Kawasaki, A.: Control of doping by matrix in few-layer graphene/metal oxide composites with highly enhanced electrical conductivity. Carbon N. Y. **81**, 83–90 (2015). https://doi.org/10.1016/j.carbon.2014.09.027
39. Huang, X., Qi, X., Boey, F., Zhang, H.: Graphene-based composites. Chem. Soc. Rev. **41**, 666–686 (2012). https://doi.org/10.1039/C1CS15078B
40. Dutkiewicz, J., Ozga, P., Maziarz, W., Pstruś, J., Kania, B., Bobrowski, P., Stolarska, J.: Microstructure and properties of bulk copper matrix composites strengthened with various kinds of graphene nanoplatelets. Mater. Sci. Eng. A. **628**, 124–134 (2015). https://doi.org/10.1016/j.msea.2015.01.018
41. Yin, S., Zhang, Z., Ekoi, E.J., Wang, J.J., Dowling, D.P., Nicolosi, V., Lupoi, R.: Novel cold spray for fabricating graphene-reinforced metal matrix composites. Mater. Lett. **196**, 172–175 (2017). https://doi.org/10.1016/j.matlet.2017.03.018
42. Dardona, S., Hoey, J., She, Y., Schmidt, W.R.: Direct write of copper-graphene composite using micro-cold spray. AIP Adv. **6** (2016). https://doi.org/10.1063/1.4961510
43. McCreary, K.M., Hanbicki, A.T., Jernigan, G.G., Culbertson, J.C., Jonker, B.T.: Synthesis of large-area WS_2 monolayers with exceptional photoluminescence. Sci. Rep. **6**, 19159 (2016). https://doi.org/10.1038/srep19159

44. Zhang, X., Xu, H., Wang, J., Ye, X., Lei, W., Xue, M., Tang, H., Li, C.: Synthesis of ultrathin ws_2 nanosheets and their tribological properties as lubricant additives. Nanoscale Res. Lett. **11**, 442 (2016). https://doi.org/10.1186/s11671-016-1659-3
45. Rengifo, S.: A comparison between graphene and WS_2 as solid lubricant additives to aluminum for automobile applications. FIU Thesis (2015)
46. Xu, S., Gao, X., Hu, M., Sun, J., Jiang, D., Wang, D., Zhou, F., Weng, L., Liu, W.: Dependence of atomic oxygen resistance and the tribological properties on microstructures of WS_2 films. Appl. Surf. Sci. **298**, 36–43 (2014). https://doi.org/10.1016/j.apsusc.2014.01.002
47. Loganathan, A., Rengifo, S., Hernandez, A.F., Emirov, Y., Zhang, C., Boesl, B., Karthikeyan, J., Agarwal, A.: Effect of 2D WS_2 addition on cold-sprayed aluminum coating. J. Therm. Spray Technol. **26**, 1585–1597 (2017). https://doi.org/10.1007/s11666-017-0608-8
48. Smid, I., Segall, A.E., Walia, P., Aggarwal, G., Eden, T.J., Potter, J.K.: Cold-sprayed Ni-hBN self-lubricating coatings. Tribol. Trans. **55**, 599–605 (2012). https://doi.org/10.1080/10402004.2012.686085
49. Neshastehriz, M., Smid, I., Segall, A.E.: In-situ agglomeration and de-agglomeration by milling of nano-engineered lubricant particulate composites for cold spray deposition. J. Therm. Spray Technol. **23**, 1191–1198 (2014). https://doi.org/10.1007/s11666-014-0122-1
50. Neshastehriz, M.: Influence of pre-process work hardening of nickel encapsulated hexagonal boron nitride powders on cold spray coatings, (2014)
51. Stark, L.M.: Engineered self-lubricating coatings utlizing cold spray technology (2010)

Chapter 8
Cold Sprayed Metallic Glass and High Entropy Alloy Deposits

Abstract Metallic glasses (MGs) and high entropy alloys (HEAs) are both emerging multi-component alloys with unique microstructures and extraordinary properties. However, there are many challenges during the manufacturing process of MGs and HEAs using conventional fabrication techniques which hinders their development as engineering materials. Cold spray as a burgeoning metal deposition and solid-state additive manufacturing technology provides an alternative to allow the fabrication of large-scale MGs and HEAs. This chapter provides an overview of MGs and HEAs deposits fabricated by cold spray additive manufacturing technique. Firstly, the background of MGs and HEAs are briefly introduced. Then, the feedstock powder for spraying and deposit microstructure evolution are provided. Finally, the particle deformation behavior and bonding mechanism are discussed.

8.1 Introduction to Metallic Glasses and High Entropy Alloys

Metallic glasses (MGs) and high entropy alloys (HEAs) are two typical advanced multi-component alloys with unique physical, mechanical and chemical properties, which have attracted wide attention in the past years and showed great potential as emerging engineering materials. MGs, also called amorphous alloys, possess long-range highly disordered and short-range ordered atomic-scale structure which can be regarded as frozen liquids with a high resistance to crystallization [1]. Compared to conventional crystalline materials, there are no defects such as grain boundaries and dislocations in MGs, which brings them superior mechanical properties (i.e., high strength, high elastic limits and high hardness), good soft magnetic properties (i.e. high saturation magnetization and high permeability) and excellent wear and corrosion resistance at room temperature [2–4]. The first reported MGs $Au_{75}Si_{25}$ was manufactured by extremely rapid cooling technique in 1960 [5]. Other manufacturing techniques of MGs, such as arc melting drop/suction casting, high-pressure die casting and copper mold casting, emerge over time which enrich the types of MGs [6]. The produce of MGs by foresaid techniques requires an extremely high cooling rate to avoid crystallization. At present, the application of MGs faces two major

S. Yin and R. Lupoi, *Cold Spray Additive Manufacturing*, Springer Tracts in Additive Manufacturing, https://doi.org/10.1007/978-3-030-73367-4_8

challenges: size limitation and poor ductility at room temperature. The required high cooling rates for retaining amorphous structure during manufacturing process limits the size of MGs. In order to break through the size limitation, three basic empirical rules are proposed: (i) the alloy must contain at least three or more components and it will be easier to form glassy structure with increasing number of components, (ii) the atomic size differences among main constituent elements should be above 12%, (iii) the heat of mixing among major constituent elements should be negative [7, 8]. Designing alloy composition based on empirical rules, bulk metallic glasses (BMGs) with amorphous structure could be formed at lower critical cooling rates. The poor ductility at room temperature is another challenge for application, which may result in catastrophic failure due to the formation of highly localized shear band during deformation [9].

HEAs are novel metal alloys which comprise multiple (typically five or more) principal elements in equimolar or near-equimolar ratio. The entropy increases with increasing the number of alloying elements in the mix; therefore, they are termed as HEAs. This new material shows excellent mechanical performance at elevated temperatures, superior ductility and fracture toughness, superconductivity and many other advantages compared to traditional metal alloys [10–12]. Current manufacturing methods for HEAs include conventional vacuum arc melting and induction melting followed by casting, where principal alloying elements have to experience remelting process to receive homogeneity microstructure [13]. High cooling rate is an indispensable prerequisite to avoid the existence of unwanted phase transformation and the formation of intermetallic phases. Consequently, HEAs suffer similar challenges of size limitation as MGs which slows down its application.

8.2 Feedstock

The feedstock powder for cold spraying is usually produced by inert gas atomization or water atomization. In general, the powders possess spherical shape with a particle size between 5 and 100 μm in diameter [14]. The materials-related parameters (i.e. particle size, flowability, surface state and mechanical properties) have a great impact on the successful particle deposition and the quality of cold spray MG and HEA deposits [15]. In the case of MG powder, mechanical alloying has also been employed for powder fabrication. No matter which production method is used, the produced feedstock is required to have a high extent of amorphous content and good glass forming ability (GFA) for high-performance deposit fabrication. For MG powder characterization, X-ray diffraction (XRD) and transmission electron microscopy (TEM) in combination with selected area electron diffraction (SAED) pattern or fast Fourier transform (FFT) pattern are usually conducted to confirm the amorphous structure of feedstock. Figure 8.1 provides an example of Cu-based MG powder characterization [16]. In addition, the thermal properties of MG play also an important role in particle deformation and deposition. Therefore, differential scanning calorimeter (DSC) is always applied for the measurement of glass transition

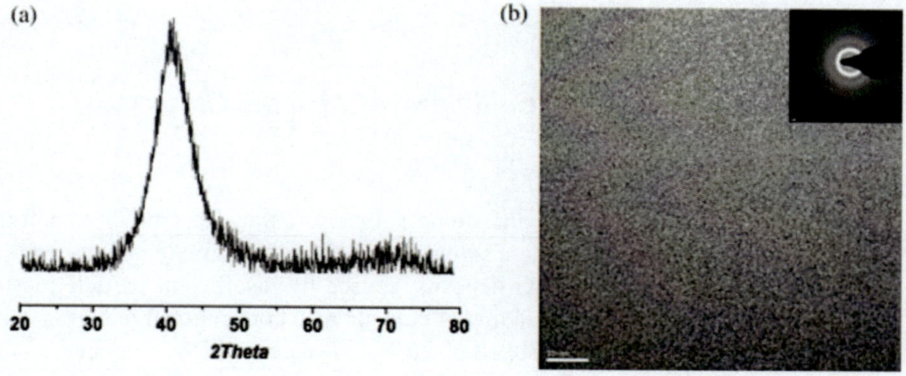

Fig. 8.1 Characteristics of the $Cu_{54}Ni_6Ti_{22}Zr_{18}$ MG feedstock: **a** XRD pattern, **b** HREM images with corresponding SAED insets [16]. Note that the broad diffraction peak in XRD pattern and diffusion halo in SAED indicate the amorphous structure in the feedstock powder [16]

temperature (T_g) and crystallization temperature (T_x), which could provide guidance for optimal manufacturing parameter selection.

In the case of the HEA feedstock for cold spraying, gas-atomized powders as shown in Fig. 8.2. are typically used, where dendritic grain structure can be clearly observed [17]. The manufacturing procedure for such HEA powders is similar to that for conventional alloy powders. Furthermore, the powder also can be synthesized via in-situ mechanical alloying. For example, Hushchyk et al. prepared AlFeNiCoCrTi HEA powder via ball milling the 5 alloying elements followed by annealing to induce solid solution and another round ball milling to remove powder agglomeration [18]. With this method, the as-produced powders may suffer difficulties in deposition as the ball milling process reduce the sparyability of the powders as discussed in Chap. 7.

Fig. 8.2 Morphology of FeCoNiCrMn HEA powder used for cold spray. **a** overall view, and **b** magnified view of a single HEA particle [17]

8.3 Microstructures

8.3.1 Microstructures of Cold Sprayed Metallic Glass Deposits

MG powders have high hardness and strength, therefore they are difficult to deform upon impact when their temperature is lower than glass transition temperature (T_g). The porosity of cold sprayed MG deposits, caused by insufficient particle plastic deformation, is usually higher than that of cold sprayed conventional metals. As the temperature is increased to the supercooled liquid region which is defined as the temperature region between T_g and crystallization temperature (T_x), thermal softening of MG particles occurs and thus they can easily deform upon impact [19]. Therefore, for MG powders, particle temperature prior to the impact is the most critical factor that determines the successful deposition. Typically, lower particle temperature tends to result in the low-quality MG deposit or even failure of deposition due to the lack of sufficient plastic deformation. Figure 8.3 shows the cross-sectional morphologies of the as-sprayed Ni-based MG deposits under different powder preheating temperatures. As can be seen, due to the lack of sufficient plastic deformation, a large number of pores and cracks can be seen in the deposit produced using non-preheated MG powders. As the preheating temperature increases, the porosity shows a significant reduction due to the improved thermal softening effect and plastic deformation of the MG powders. Together with the improved density, the deposition efficiency also increases dramatically with increasing the powder preheating temperature, which suggests an enhanced interparticle bond [20]. However, increasing powder preheating temperature brings negative aspects to the deposit at the same time. Oxidization and crystallization are more likely to occur when the powder preheating temperature is at a high level. Figure 8.4 shows the content of oxygen and amorphous phase of Ni-based MG deposit as function of powder preheating temperature. An increasing trend can be identified for oxygen, and the content of amorphous phase reduces. The decreased amorphous content was also found in cold sprayed Al-based MG deposits when propulsive gas temperature was increased from 400 to 600 °C [21], which further validate the critical role of particle temperature in the MG deposition. Therefore, during the deposition of MG powders, it is necessary

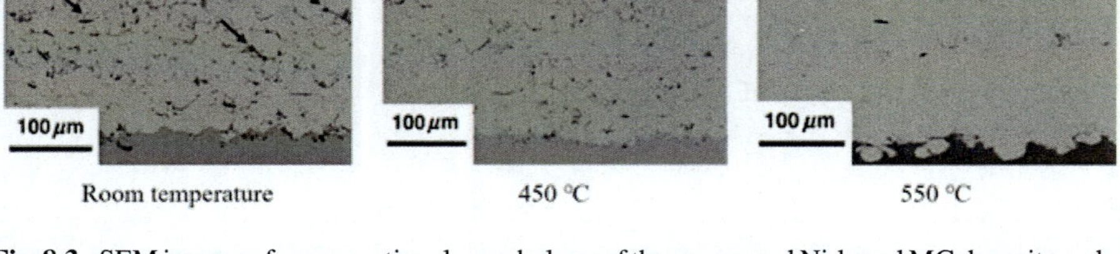

Fig. 8.3 SEM images of cross-sectional morphology of the as-sprayed Ni-based MG deposits under different powder preheating temperature, **a** room temperature, **b** 450 °C, **c** 550 °C [22]

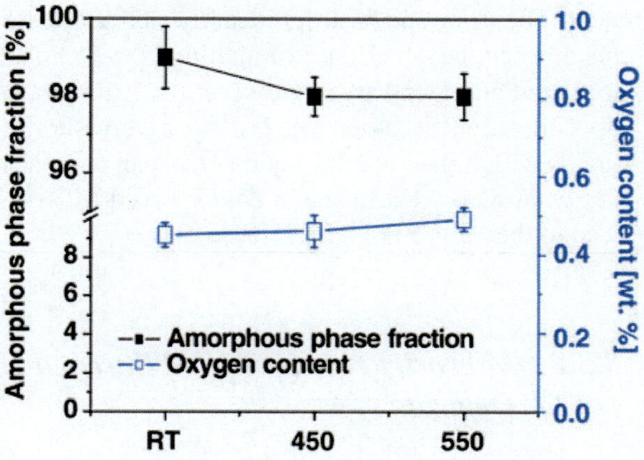

Fig. 8.4 Effects of powder preheating temperature on the amorphous phase fraction and oxygen content of the as-sprayed Ni-based MG deposit [22]

to consider the balance between high energetic conditions to induce sufficient plastic deformation and low thermal energy to avoid oxidation and crystallization.

Crystallization of MG particles during cold spray deposition is not an unusual phenomenon. Several studies have reported the existence of nano-sized crystals at the highly deformed regions in MG deposits (i.e. interparticle interfaces where MG experiences severe plastic deformation and considerable temperature rise [16, 23, 24]. Figure 8.5 shows the nanocrystalline phases found at the interparticle interfaces in Zr-based MG deposit. The formation mechanism of the nanocrystalline grains can be summarized as follow. When MG particle impacts onto the substrate with high velocity, the particles undergo extensive plastic deformation at extremely high strain

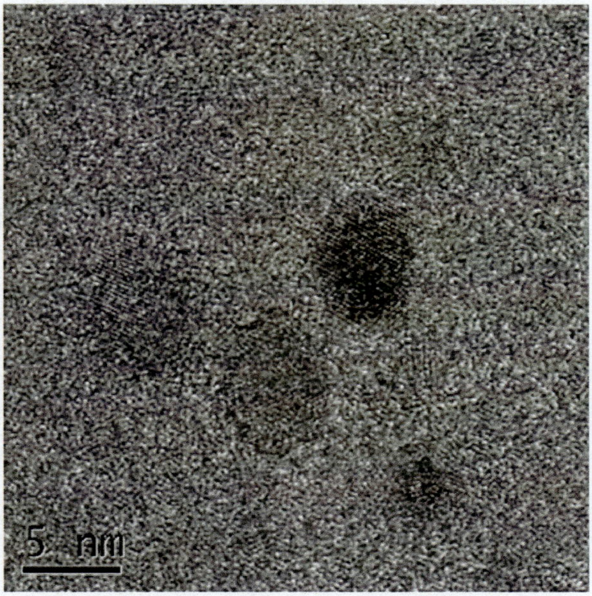

Fig. 8.5 Nanocrystalline phases formed at interparticle interface in $Zr_{55}Cu_{30}Ni_5Al_{10}$ MG deposit which was fabricated in the author's laboratory using nitrogen as propelling gas at 3.0 MPa and 800 °C

rates. The strain and heat significantly increase at the localized interparticle interface, resulting in the occurrence of melting ($T > T_g$). As a consequence, crystals form at localized interfacial areas. However, as cooling rate is extremely high (10^{-8} s), such crystallization behavior can last for a very short period and only nanocrystalline grains with a size of a few nanometers can form near the interface. Typically, this is not what we expect to see in cold sprayed MG deposits as the crystallization may lead to the loss of properties.

8.3.2 Microstructures of Cold Sprayed High Entropy Alloy Deposits

Currently, the number of studies on cold sprayed HEA deposits are very limited. Table 8.1 summarizes the works with regard to cold sprayed HEA available in literature. The authors' group is the very first research team worldwide that confirms the feasibility of cold spray for fabricating HEA deposits [17]. As shown in Fig. 8.6, the cold sprayed FeCoNiCrMn deposit presents a dense structure without visible pores and cracks and has homogeneous element distribution. From the EBSD and XRD results shown in Fig. 8.7, the grain structure of the deposit was significantly refined compared to that of feedstock powder due to the increased dislocation density and dynamic recrystallization during the deposition process. It can be observed from XRD spectra that the cold sprayed HEA deposit retained original phases as in the feedstock powders. However, interparticle interfaces as pointed by arrows shown in Fig. 8.6a can be found in the cold sprayed HEA deposit, which is attributed to the insufficient local plastic deformation. The occurrence of interparticle interfaces in cold spray may lead to a decreased bonding strength and mechanical properties.

In order to heal the interparticle interfaces and improve the mechanical properties of as-fabricated HEA deposit, post heat-treatment can be used. Figure 8.8 shows the microstructure of CoCrFeMnNi HEA deposits before and after heat treatment [25]. It is clear that the grains in cold sprayed HEA deposit presents heterogeneous structure, which is quite similar to that of cold sprayed conventional metals. Coarse grains are present at the central region of each particle, and ultra-fine grains can be found at the interparticle interfaces, which could be attributed to the localized severe plastic deformation and the resultant dynamic recrystallization [17, 25]. After heat treatment at 600 °C for 1 h, the grains grow into coarse ones due to the occurrence of recrystallization during heat treatment. The ultra-fine grains at the interparticle interfaces also coarsen as marked by white arrows in Fig. 8.8b but are still finer than the grains in the particles. A recent study applied higher heat treatment temperature to cold sprayed CoCrFeMnNi deposit, and twins are formed in the deposit [26].

Table 8.1 Summarization of published works with regard to cold sprayed HEA deposits

Feedstock	Substrate	Gas parameters			Heat treatment	Thickness (mm)	Porosity (%)	Hardness	Ref
		Type	Pressure (MPa)	Temperature (°C)					
FeCoNiCrMn (15 ~ 53 μm)	6082 Al alloy	He	3.0	300	-	1.5	0.47 ± 0.17	332.91 ± 34.74 $HV_{0.1}$	[17]
AlCoCrFeNi (d_{50} = 7 μm)	Ni-based superalloy	compressed air	1.0	400	1100 °C for 25 h	-	-	-	[27]
CrFeNiMn (20 ~ 45 μm)	Fe52 steel	N_2	5.0–6.0	-	-	0.23–0.49	3–10	> 300 $HV_{0.3}$	[28]
CoCrFeMnNi (10 ~ 60 μm)	steel	He	3.0	300	As sprayed	1	0.32	423.8 $HV_{0.5}$	[26]
					550 °C for 2 h	–	–	399.6 $HV_{0.5}$	
					850 °C for 2 h	–	–	219.9 $HV_{0.5}$	
FeCoCrNiMn (15 ~ 53 μm)	6082 Al alloy	He	3.0	300	As-sprayed	>3	~ 0.5	–	[25]
					600 °C for 1 h		–	–	
AlNiCoFeCrTi (1 ~ 125 μm)	Low carbon steel	air	0.9	450	–	~ 0.45	<1	1030 ± 30 $HV_{0.1}$	[18]

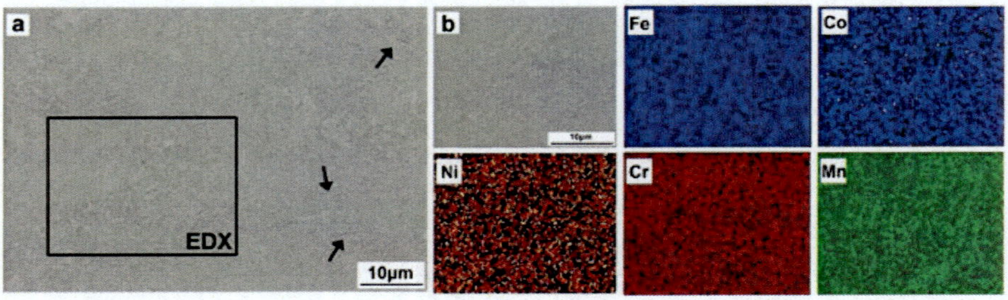

Fig. 8.6 **a** Cross-sectional morphology of cold sprayed FeCoNiCrMn HEA deposit and **b** EDX map [17]

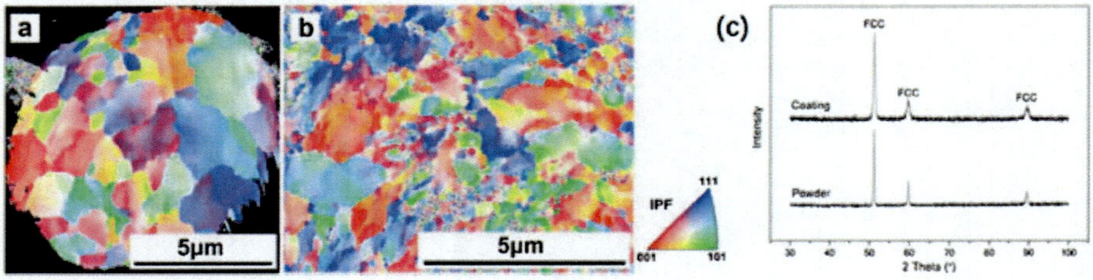

Fig. 8.7 EBSD IPF maps of **a** a single HEA particle and **b** the cold sprayed HEA deposit; **c** XRD spectra of powder and deposit [17]

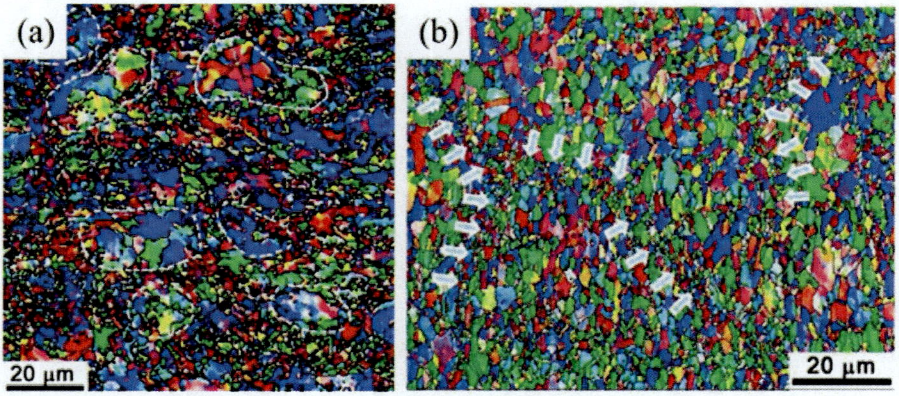

Fig. 8.8 EBSD IPF maps of equiatomic FeCoCrNiMn HEA deposit. **a** as-sprayed, **b** heat treatment [25]

8.4 Bonding Mechanism of Cold Sprayed Metallic Glass Deposits

The plastic deformation of MG, which depends heavily on the temperature and strain rate, can be categorized into inhomogeneous deformation (localized flow) and homogeneous deformation (plastic flow) [4, 29]. Figure 8.9 shows the schematic deformation map which can be used to predict the deformation of MG particle during cold spray process. The MG particle experiences inhomogeneous deformation from room temperature to T_g, which can be characterized by severe localized plastic deformation at the contact region and fracture [30]. The localized flow of MG particle exists even when the temperature is increased above T_g, which could be attributed to the extreme high strain rate in cold spray (from 10^6 to 10^8 s^{-1}). From the map, it can be found that increasing particle temperature or reducing strain rate pushes the particle from inhomogeneous deformation to homogeneous deformation. Typically, homogeneous deformation facilitates particle deposition and deposit growth. However, elevated temperature also leads to inevitable nucleation and growth of crystalline phases, which reduces the purity of amorphous phases. Figure 8.10 shows the particle morphologies of single Zr-based MG particle deposited onto Zr-based MG substrate without particle preheating. It can be found that the non-preheated particle mostly remains solid state and significantly fractures. This is a typical inhomogenous deformation for cold sprayed MG particles, which will be discussed in the next paragraph. In contrast, for the particle preheated to a temperature between T_g and T_x, a clear

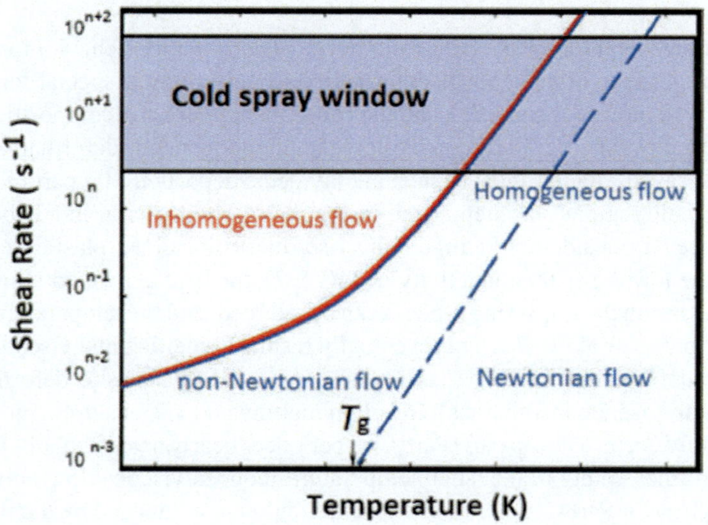

Fig. 8.9 Schematic deformation map for MG particles in cold spray process. The window of shear rates depends on spraying conditions (standoff distance, substrate hardness, gas pressure, and gas temperature) [31]

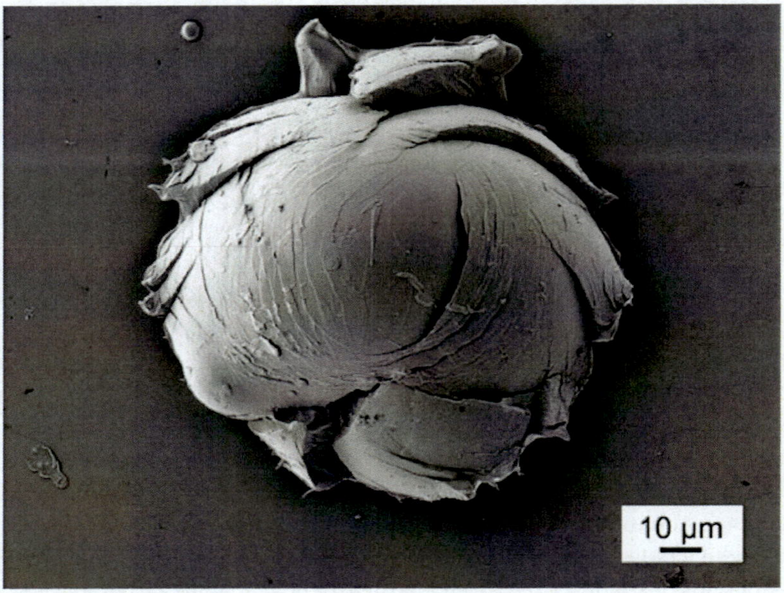

Fig. 8.10 Top view of the morphologies of a Zr-based MG particle depositing onto Zr-based MG substrate. The experiment was implemented in the author's laboratory using nitrogen as propelling gas at 3.0 MPa and 800 °C

sign of the occurrence of viscus flow can be observed and thus its deformation is homogenous [24].

Inhomogeneous deformation is a more typical deformation behavior for cold spray MG particles than homogeneous deformation as cold spray is special for its unique capability to prevent the melting and the consequent phase and composition changes of the particles. In order to further explore the inhomogeneous deformation behavior of cold sprayed MG particle, Yoon et al., divided a deposited MG particle into three characteristic regions as schematized in Fig. 8.11. The material in adiabatic region experiences a considerable temperature rise due to localized plastic deformation. Due to the low thermal conductivity of MG [32], the heat generated at interparticle interface during the impacting process can be retained, and the temperature can reach a level close to or above T_g. This eventually results in melting and viscous flow, and thus the deformation in Part C is severe. At part B, the plastic deformation and temperature rise are insufficient to result in melting and viscous flow, and thus shear and fracture occur. The top part (Part A) experiences nearly no deformation. However, studies further point out that such temperature-induced viscous flow only occurs at the periphery of Part C rather than the entire region as evidenced by the pronounced vain pattern at the bottom of a detached Cu-based MG particle shown in Fig. 8.12. This phenomenon may also suggest that deposition occurs when viscous force of the liquefied interface surpasses the particle rebounding force. Timely solidification is also required to prevent particle lift-off.

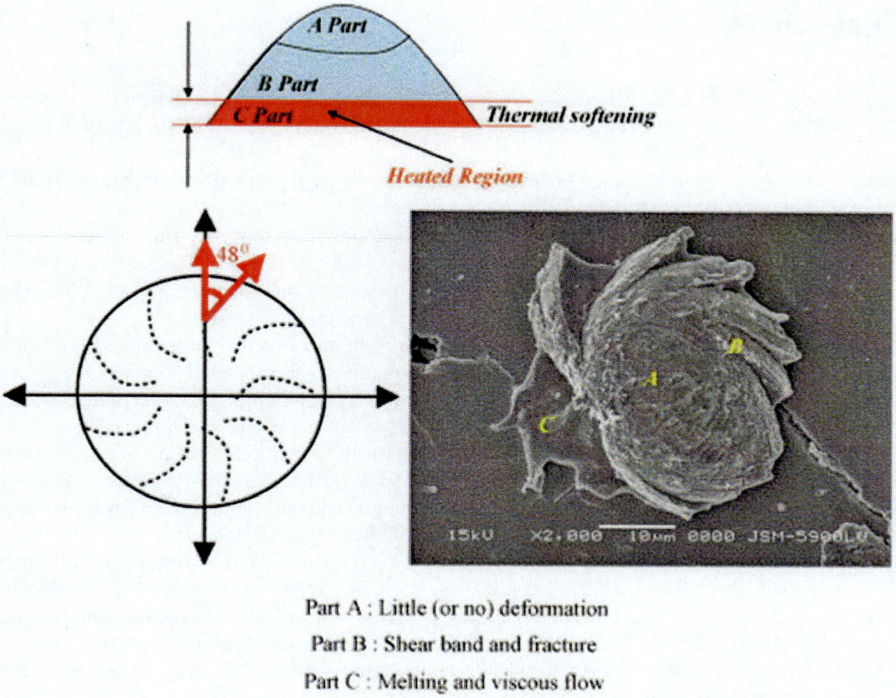

Fig. 8.11 Deposition behavior of NiTiZrSiSn BMG particles [30]

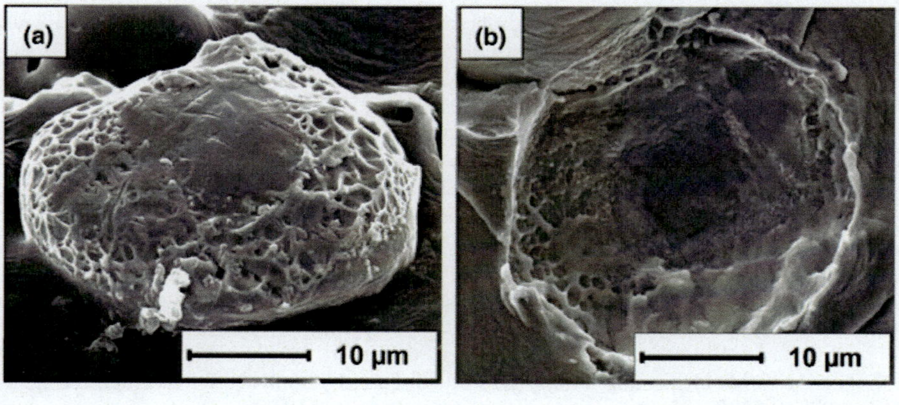

Fig. 8.12 **a** Single $Cu_{50}Zr_{50}$ particle detached after cavitation testing and **b** a corresponding crater on the substrate surface, both showing pronounced vain pattern [33]

References

1. Pauly, S., Löber, L., Petters, R., Stoica, M., Scudino, S., Kühn, U., Eckert, J.: Processing metallic glasses by selective laser melting. Mater. Today. **16**, 37–41 (2013). https://doi.org/10.1016/j.mattod.2013.01.018
2. Wang, W.H.: Bulk metallic glasses with functional physical properties. Adv. Mater. **21**, 4524–4544 (2009). https://doi.org/10.1002/adma.200901053
3. Kruzic, J.J.: Bulk metallic glasses as structural materials: a review. Adv. Eng. Mater. **18**, 1308–1331 (2016). https://doi.org/10.1002/adem.201600066
4. Trexler, M.M., Thadhani, N.N.: Mechanical properties of bulk metallic glasses. Prog. Mater. Sci. **55**, 759–839 (2010). https://doi.org/10.1016/j.pmatsci.2010.04.002
5. Klement, W., Willens, R.H., Duwez, P.: Non-crystalline structure in solidified gold-silicon alloys. Nature **187**, 869–870 (1960). https://doi.org/10.1038/187869b0
6. Bakkal, M., Karagüzel, U., Kuzu, A.T.: Manufacturing techniques of bulk metallic glasses. In: Modern Manufacturing Processes. pp. 137–148. Wiley, Hoboken, NJ, USA (2019)
7. Inoue, A.: High strength bulk amorphous alloys with low critical cooling rates (Overview). Mater. Trans. JIM. **36**, 866–875 (1995). https://doi.org/10.2320/matertrans1989.36.866
8. Inoue, A.: Stabilization of metallic supercooled liquid and bulk amorphous alloys. Acta Mater. **48**, 279–306 (2000). https://doi.org/10.1016/S1359-6454(99)00300-6
9. Zhang, Z.F., He, G., Eckert, J., Schultz, L.: Fracture mechanisms in bulk metallic glassy materials. Phys. Rev. Lett. **91**, 2–5 (2003). https://doi.org/10.1103/PhysRevLett.91.045505
10. Ye, Y.F., Wang, Q., Lu, J., Liu, C.T., Yang, Y.: High-entropy alloy: challenges and prospects. Mater. Today. **19**, 349–362 (2016). https://doi.org/10.1016/j.mattod.2015.11.026
11. George, E.P., Raabe, D., Ritchie, R.O.: High-entropy alloys. Nat. Rev. Mater. **4**, 515–534 (2019). https://doi.org/10.1038/s41578-019-0121-4
12. Zhang, Y., Zuo, T.T., Tang, Z., Gao, M.C., Dahmen, K.A., Liaw, P.K., Lu, Z.P.: Microstructures and properties of high-entropy alloys. Prog. Mater. Sci. **61**, 1–93 (2014). https://doi.org/10.1016/j.pmatsci.2013.10.001
13. Li, X.: Additive manufacturing of advanced multi-component alloys: bulk metallic glasses and high entropy alloys. Adv. Eng. Mater. **1700874**, 1700874 (2017). https://doi.org/10.1002/adem.201700874
14. Yin, S., Cavaliere, P., Aldwell, B., Jenkins, R., Liao, H., Li, W., Lupoi, R.: Cold spray additive manufacturing and repair: Fundamentals and applications. Addit. Manuf. **21**, 628–650 (2018). https://doi.org/10.1016/j.addma.2018.04.017
15. Jodoin, B., Ajdelsztajn, L., Sansoucy, E., Zúñiga, A., Richer, P., Lavernia, E.J.: Effect of particle size, morphology, and hardness on cold gas dynamic sprayed aluminum alloy coatings. Surf. Coatings Technol. **201**, 3422–3429 (2006). https://doi.org/10.1016/j.surfcoat.2006.07.232
16. Yoon, S., Bae, G., Xiong, Y., Kumar, S., Kang, K., Kim, J.J., Lee, C.: Strain-enhanced nanocrystallization of a CuNiTiZr bulk metallic glass coating by a kinetic spraying process. Acta Mater. **57**, 6191–6199 (2009). https://doi.org/10.1016/j.actamat.2009.08.045
17. Yin, S., Li, W., Song, B., Yan, X., Kuang, M., Xu, Y., Wen, K., Lupoi, R.: Deposition of FeCoNiCrMn high entropy alloy (HEA) coating via cold spraying. J. Mater. Sci. Technol. **35**, 1003–1007 (2019). https://doi.org/10.1016/j.jmst.2018.12.015
18. Hushchyk, D.V., Yurkova, A.I., Cherniavsky, V.V., Bilyk, I.I., Nakonechnyy, S.O.: Nanostructured AlNiCoFeCrTi high-entropy coating performed by cold spray. Appl. Nanosci. (2020). https://doi.org/10.1007/s13204-020-01364-4
19. Choi, H., Yoon, S., Kim, G., Jo, H., Lee, C.: Phase evolutions of bulk amorphous NiTiZrSiSn feedstock during thermal and kinetic spraying processes. Scr. Mater. **53**, 125–130 (2005). https://doi.org/10.1016/j.scriptamat.2005.01.046
20. Yoon, S., Lee, C., Choi, H., Kim, H., Bae, J.: Impacting behavior of bulk metallic glass powder at an abnormally high strain rate during kinetic spraying. Mater. Sci. Eng. A. **448–451**, 911–915 (2007). https://doi.org/10.1016/j.msea.2006.02.433

21. Jin, L., Che, Z., Zhang, L., Zhang, B., Fan, Z., Chen, Y.: The composition design and properties investigation of Al-based amorphous powder. Int. J. Light. Mater. Manuf. **3**, 198–203 (2020). https://doi.org/10.1016/j.ijlmm.2019.12.001
22. Yoon, S., Kim, H.J., Lee, C.: Deposition behavior of bulk amorphous NiTiZrSiSn according to the kinetic and thermal energy levels in the kinetic spraying process. Surf. Coatings Technol. **200**, 6022–6029 (2006). https://doi.org/10.1016/j.surfcoat.2005.09.022
23. Yoon, S., Xiong, Y., Kang, K., Bae, G., Lee, C.: Phase separation in kinetic sprayed bulk metallic glasses. J. Phys. D. Appl. Phys. **42**, 182007 (2009). https://doi.org/10.1088/0022-3727/42/18/182007
24. Yoon, S., Xiong, Y., Kim, H., Lee, C.: Dependence of initial powder temperature on impact behaviour of bulk metallic glass in a kinetic spray process. J. Phys. D. Appl. Phys. **42**, 082004 (2009). https://doi.org/10.1088/0022-3727/42/8/082004
25. Xu, Y., Li, W., Qu, L., Yang, X., Song, B., Lupoi, R., Yin, S.: Solid-state cold spraying of FeCoCrNiMn high-entropy alloy: an insight into microstructure evolution and oxidation behavior at 700–900 °C. J. Mater. Sci. Technol. **68**, 172–183 (2020). https://doi.org/10.1016/j.jmst.2020.06.041
26. Ahn, J.-E., Kim, Y.-K., Yoon, S.-H., Lee, K.-A.: Tuning the microstructure and mechanical properties of cold sprayed equiatomic CoCrFeMnNi high-entropy alloy coating layer. Met. Mater. Int. (2020). https://doi.org/10.1007/s12540-020-00886-4
27. Anupam, A., Kumar, S., Chavan, N.M., Murty, B.S., Kottada, R.S.: First report on cold-sprayed AlCoCrFeNi high-entropy alloy and its isothermal oxidation. J. Mater. Res. **34**, 796–806 (2019). https://doi.org/10.1557/jmr.2019.38
28. Lehtonen, J., Koivuluoto, H., Ge, Y., Juselius, A., Hannula, S.P.: Cold gas spraying of a high-entropy CrFeNiMn equiatomic alloy. Coatings. **10**, (2020). https://doi.org/https://doi.org/10.3390/coatings10010053
29. Henao, J., Concustell, A., Cano, I.G., Cinca, N., Dosta, S., Guilemany, J.M.: Influence of Cold Gas Spray process conditions on the microstructure of Fe-based amorphous coatings. J. Alloys Compd. **622**, 995–999 (2015). https://doi.org/10.1016/j.jallcom.2014.11.037
30. Yoon, S., Lee, C., Choi, H., Jo, H.: Kinetic spraying deposition behavior of bulk amorphous NiTiZrSiSn feedstock. Mater. Sci. Eng. A. **415**, 45–52 (2006). https://doi.org/10.1016/j.msea.2005.08.132
31. Henao, J., Sharma, M.M.: Characterization, deposition mechanisms, and modeling of metallic glass powders for cold spray. In: Cold-Spray Coatings. pp. 251–272. Springer International Publishing, Cham (2018)
32. Yamasaki, M., Kagao, S., Kawamura, Y.: Thermal diffusivity and conductivity of Zr55Al10Ni5Cu30 bulk metallic glass. Scr. Mater. **53**, 63–67 (2005). https://doi.org/10.1016/j.scriptamat.2005.03.021
33. List, A., Gärtner, F., Mori, T., Schulze, M., Assadi, H., Kuroda, S., Klassen, T.: Cold spraying of amorphous Cu50Zr50 alloys. J. Therm. Spray Technol. **24**, 108–118 (2014). https://doi.org/10.1007/s11666-014-0187-x

Chapter 9
Industrial Applications of Cold Spray Additive Manufacturing

Abstract CSAM as an emerging additive manufacturing process has great potentials in fabricating free-stranding components and repairing damage components. It can be used in a wide range of industrial sectors such as aerospace, aviation, marine, automotive and nuclear. Although CSAM has been regarded as an additive manufacturing process in recent years, it typically leaves a rough, undulating and porous surface after fabrication. For practical industrial applications, such as in the repair of a worn surface, the as-fabricated surface is not suitable for immediate use, either due to surface finish or dimensional accuracy requirements. This necessitates the use of a machining or finishing process. For clarifying the machinability of CSAM deposits and explore their potential applications, this chapter provides a thorough review and discussion on the post machining and the industrial applications of CSAM deposits.

9.1 Machining of CSAM Deposits

9.1.1 Machinability of CSAM Deposits

In general, CSAM products are suitable for normal machining processes. As examples, Fig. 9.1 shows three CSAM products made using a high pressure cold spray system located in Trinity College Dublin (TCD), which were machined into finished parts using normal machining processes. The aluminium alloy tube shown in Fig. 9.1a was finished by turning; the aluminium alloy flange shown in Fig. 9.1b was finished by turning and drilling; the copper cuboid shown in Fig. 9.1c was finished by milling. These cases clearly demonstrate the suitability of normal machining processes for CSAM products.

When understanding the machining of CSAM deposits, it is important to recognize that unless the deposit has been brought to full density or high ductility, through proper annealing, hot isostatic pressing (HIP) or a similar method, it will machine similarly to a porous or cast material. As CSAM deposits are an amalgamation of individual particles, there will be variations in density and levels of work hardening within the deposit, with the potential for small oxide inclusions [1]. This will cause variations in the mechanical and thermal loads experienced by the cutting tool. Many

S. Yin and R. Lupoi, *Cold Spray Additive Manufacturing*, Springer Tracts in Additive Manufacturing, https://doi.org/10.1007/978-3-030-73367-4_9

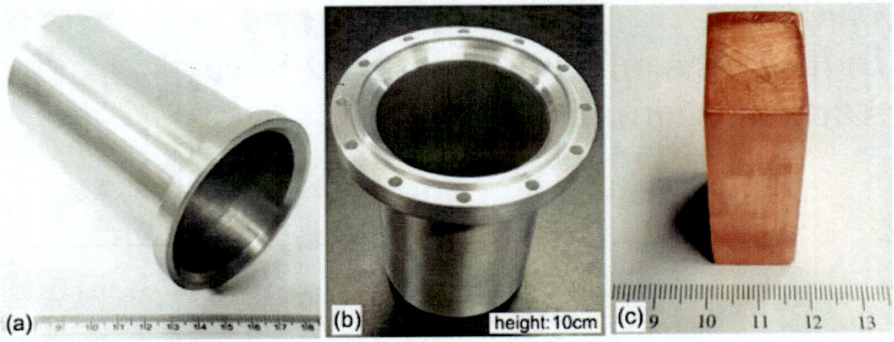

Fig. 9.1 CSAM tube machined into a finished part fabricated in TCD. **a** aluminium alloy tube, **b** aluminium alloy flange, and **c** copper cuboid

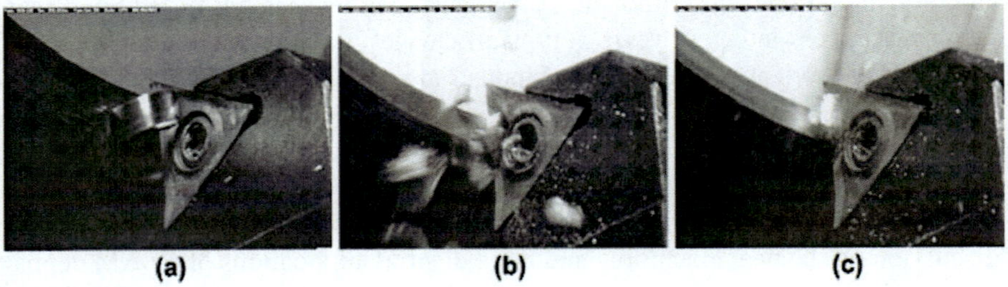

Fig. 9.2 Comparison of the chip formation mechanisms between bulk and CSAM aluminium with and without annealing during post-machining. **a** bulk material, **b** annealed CSAM deposit and **c** non-annealed CSAM deposit [6]

comparisons can be made between the machining of CSAM deposits and materials produced through powder metallurgy [2] and fusion-based additive manufacturing [3–5], which are acknowledged as more problematic to machine when compared to bulk materials. Figure 9.2 shows a comparison of the chip formation mechanisms between bulk aluminium and CSAM aluminium with and without annealing during machining [6]. Fully dense bulk aluminium generally results in continuous chip formation during machining as shown in Fig. 9.2a. Annealed CSAM aluminium which has good ductility forms discontinuous chips, but otherwise machines similar to the bulk material (Fig. 9.2b). However, for the as-fabricated CSAM aluminium as shown in Fig. 9.2c, due to the poor inter-particle bonding and little particle deformation, the cutting process resembles that of a sintered material, with inter-particle bonds being broken and no chips being formed. The surface finish after machining of any such material would be porous and undulating, though some smearing and densification of the surface may be observed [7]. Thus, the machinability of a CSAM product is dependent upon the density and ductility of the deposit created.

Other issues which are of interest in the machinability of CSAM product are the potential for debonding of the deposits if overly aggressive cutting parameters are used, or if a significant portion of the bond area is removed during machining. If it is necessary to remove the deposits from the substrate to achieve the geometry required, this should be carried out first, as the CSAM product may distort due to residual stresses after debonding.

9.1.2 Cutting Forces During the Machining of CSAM Deposits

Cutting forces during the machining of porous CSAM deposits will typically be lower than that of bulk material, but with more variation during cutting [6]. Porous materials manufactured using other techniques typically show lower but more variable cutting forces [8]. The lower magnitude of the forces is at least partially attributable to the lower stiffness of porous materials [9], while the variation is due to the partially discontinuous cutting action experienced as the tool disengages and then re-engages when passing through pores. CSAM deposits with higher ductility will show less variation in cutting forces, while deposits with lower porosity will show higher cutting forces.

9.1.3 Tool Wear During the Machining of CSAM Deposits

Little to no research has been carried out into tool wear during the machining of CSAM deposits. It can be expected that, due to the discontinuous nature of the machining process, tool wear will be greater than that observed when machining bulk material. The use of tamping materials in the CSAM process such as alumina, which are typically very hard, will cause excessive tool wear during machining [10]. This has been observed by the authors in the preparation by milling of tensile test samples from CSAM diamond-reinforced metal matrix composite. 5–10 tools were required per sample, with tool life being measured in seconds. Further research work is required to characterize the tool wear mechanisms in CSAM materials.

9.1.4 Properties and Microstructure of CSAM Deposits After Machining

Turning is the primary machining operation for the finishing of cylindrical CSAM deposits. For as-fabricated CSAM deposits, turning may not be suitable for removal of all excess material. Some deposits will fail at inter-particle boundaries and separate from the deposit or substrate as a small clump of particles, due to weak inter-particle bonding. As a consequence, the surface of as-fabricated CSAM deposits after turning may have a smeared surface with high surface roughness, as shown in Fig. 9.3a [6].

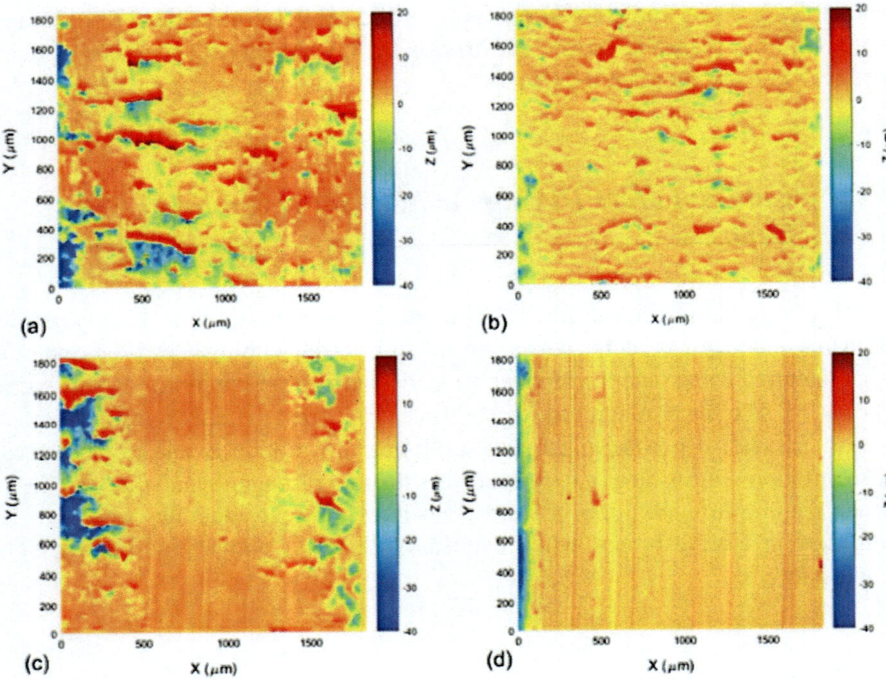

Fig. 9.3 White light interferometer maps of turned surfaces of aluminium deposits. **a** as-fabricated deposit produced by nitrogen, **b** heat-treated deposit produced by nitrogen, **c** as-fabricated deposit produced by helium, and **d** heat-treated deposit produced by helium [6]

By increasing particle impact velocity or using a heat treatment process prior to machining, the surface roughness of the turned CSAM deposits can be significantly reduced due to the improved inter-particle bonding, as shown in Fig. 9.3b–d [6]. In addition, turning also removes the compressive residual stress of the as-fabricated CSAM deposits and introduces tensile stress to machined surface of the deposits [11].

Milling is required for the finishing of CSAM deposits without axial symmetry. In recent work by the authors, milling was found to influence the deposit microstructure close to the machined surface. Figure 9.4 shows the cross-sectional image of a CSAM aluminium deposit after milling. Clearly, the milling process compresses the top layer of the deposit, making it much denser than the original deposit. In addition, milling tends to break the brittle CSAM deposit due to the large cutting force. Improper milling parameters will cause cracks, fracture and delamination of the CSAM deposit [11, 12]. Therefore, careful selection of machining parameters is required to achieve a satisfactory finish. Careful inspection for cracks and damage should be carried out on the CSAM product after machining.

Currently, investigations on the properties and microstructure of CSAM deposits after machining are still very limited. Therefore, more studies on this topic are

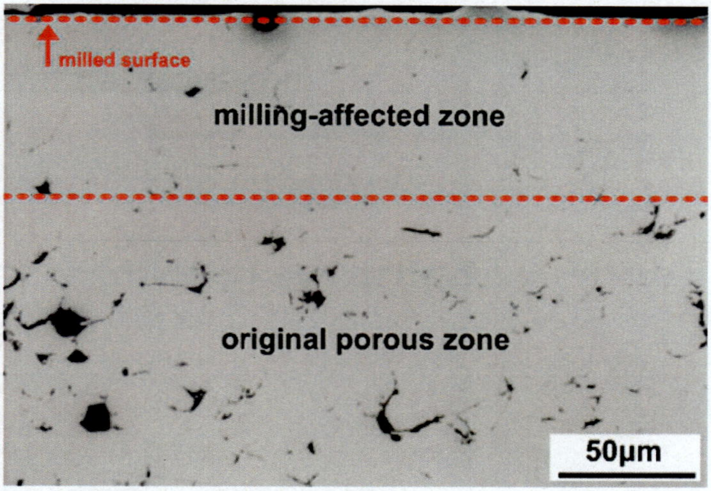

Fig. 9.4 Cross-sectional image of a CSAM pure aluminium deposit after milling. The aluminium deposit was fabricated using nitrogen with the pressure of 3 MPa and temperature of 400 °C by the cold spray system in TCD

needed in the future to fully clarify the effect of machining methods and machining parameters on the grain structure, microstructure and properties of CSAM deposits.

9.2 CSAM for Fabricating Free-Standing Components

9.2.1 Rotational Structures

A broad range of components have rotational structures (e.g. tubes, flanges and cylinder walls). CSAM has the capability to fabricate such components. Using an external spindle to hold the mandrel substrate, the rotational structure can be easily fabricated through a simple spray strategy. For instance, as shown in Fig. 9.1b, the aluminium flange was successfully fabricated for an Irish company (Moog) in the authors' lab in TCD. Aluminium was first deposited onto an aluminium tube which was held and rotated by a spindle, then the as-fabricated raw product was post-machined using turning and drilling (including a cut which crosses the interface between the substrate and CSAM deposit) to finalise the flange feature. Tests of the mechanical properties have demonstrated that this method of manufacture completely meets the requirements of the company. A similar CSAM copper flange is also shown in Fig. 9.5 [13]. Here, it should be noted that CSAM is only able to produce a rough structure rather than net-shape structure. Thus the as-fabricated CSAM deposits normally require a post-machining (subtractive manufacturing) process to produce

Fig. 9.5 Digital photos of a CSAM copper flange [13]

the required shape. This is not only applicable to rotational structures, but also to all CSAM products.

Metal cylinder outer and inner walls can also be produced using CSAM [14–20]. The fabrication process is similar to the flange fabrication, particularly for the case of an outer wall. Figure 9.6a provides an example showing a 1/10 scaled

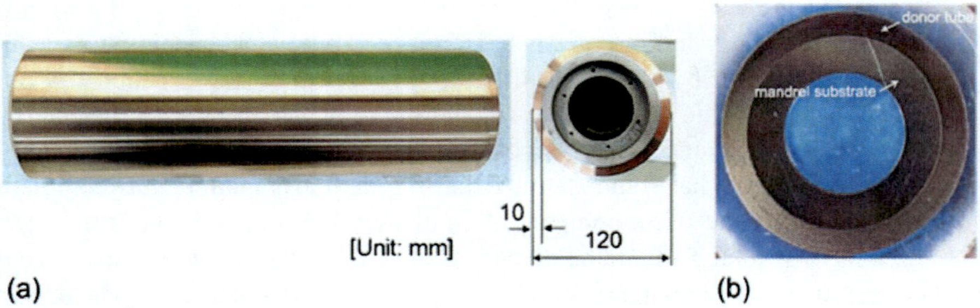

Fig. 9.6 Digital photos of CSAM rotational structure: **a** scaled canister for disposal of CANDU spent fuels and **b** tantalum-10 tungsten alloy donor tube used for the gun barrel liners [14, 15]

canister for disposal of CANDU spent fuels, fabricated using CSAM [15]. A 10 mm copper layer with a porosity of 0.3%, density of 8900 kg/m^3 and oxygen content of 0.019% (compared to 0.02% oxygen content of the feedstock copper powder) was deposited onto the cast iron cylinder, with good tensile strength, mechanical stability and thermal properties. For certain components, the mandrel substrate is not needed and has to be removed during the post-machining process. An instance of this kind of product is provided in Fig. 9.6b, showing a CSAM tantalum-10 tungsten alloy donor tube used for the fabrication of gun barrel liners [14]. The tantalum-10 tungsten alloy layer was deposited onto a cylindrical aluminium mandrel substrate, after which the inner mandrel was removed using deep drilling, honing, and a brief soak in dilute sodium hydroxide.

For the inner wall fabrication, the manufacturing procedure varies according to the inner diameter. If the inner diameter is sufficiently large that the conventional cold spray nozzle and its holding accessories can be placed within the component, the fabrication process is same as producing the outer wall. But in circumstances where the component is not large enough to accommodate a cold spray nozzle, the nozzle has to be slightly inclined to fit the limited space, at the expense of lower deposit quality due to the spray angle. An example is provided in Fig. 9.7a, showing a copper inner wall of a pressure ring from a food processing machine, which was fabricated using CSAM [20]. In this case, the nozzle had to be inclined to the target substrate during the fabrication due to the limitation of the inside diameter of the component. Furthermore, if the inner diameter is too small to accommodate a normal cold spray nozzle, a specially-designed short cold spray nozzle can be used [19, 21]. Figure 9.7b shows a short cold spray nozzle designed for small-diameter inner walls. The nozzle has a length of only 6.5 mm, far less than a normal cold spray nozzle. Figure 9.7c shows a copper inner wall on a metal tube with the diameter of 80 mm fabricated using this short nozzle [19].

CSAM is also capable of producing other special rotational structures such as cones and gears, as shown in Fig. 9.8 [19, 22]. To fabricate such structures, an elaborate design and scanning trajectory must be developed before manufacture.

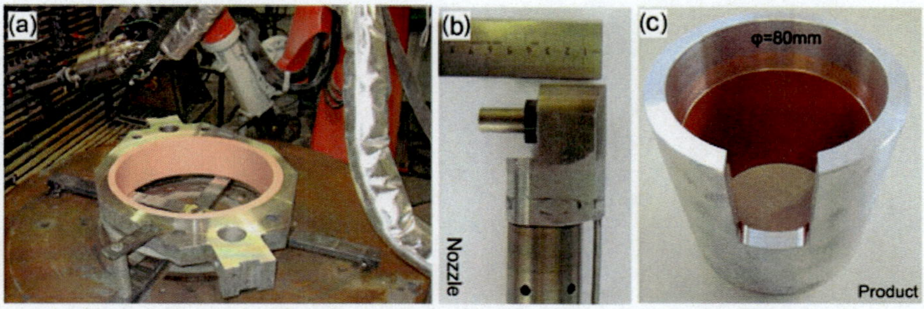

Fig. 9.7 Digital photos of CSAM rotational structure: **a** an inner wall of a pressure ring for food processing machine, **b** a small-size cold spray nozzle and **c** an inner wall of small-space cylinder tube [19, 20]

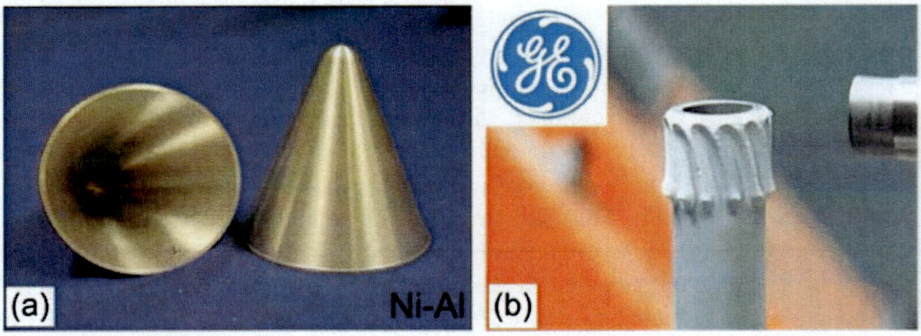

Fig. 9.8 Digital photos of CSAM components with complex structures: **a** cone structure, and **b** gear [19, 22]

Nozzle traverse speed, scanning step and trajectory should be linked to the spindle axis. There are very few studies on such special rotational structures, and therefore this topic may be a future research focus.

9.2.2 Complex Structures

A combination of CSAM and conventional subtractive manufacturing has recently shown great potential in fabricating components with complex structures. Figure 9.9 shows a component with a complex structure which was manufactured using CSAM [23]. To fabricate this component, the additive material was first deposited onto a specially designed substrate. After spraying the substrate was removed and only the CSAM part remained. Through a combination of CSAM and subtractive manufacturing, a complex component was produced. The challenge of this process is the

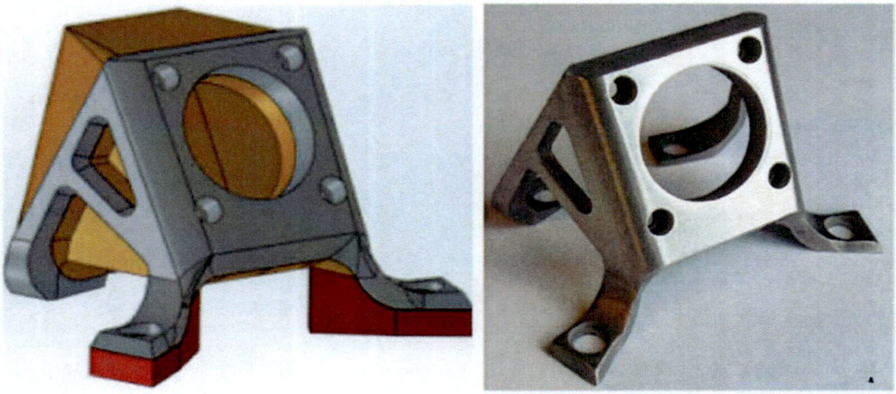

Fig. 9.9 Component with complex structure made by CSAM [23]

design of the substrate to ensure ease of removal, and to minimize post-process machining. Based on this technique CSAM can also be used to modify an existing component. Figure 9.10 shows the modification process of adding a new part onto a bearing cap via CSAM. As can be seen, a completely new component was obtained after CSAM. Visually, there is no obvious boundary between the new part and original component. The new component passed the customer's quality and performance specifications [24], proving the feasibility of CSAM for this application. Fabrication of such complex structures is very important to CSAM technology as this can significantly broaden the application areas. Therefore, further investigation is suggested in this area of CSAM.

CSAM can also produce array structures. As shown in Fig. 9.11a, the fabrication process needs a specially designed mask to prevent the deposition of redundant materials to allow only the deposition of the desired pattern. An example of a CSAM array structure is the pyramidal fin arrays used for a compact heat exchanger, as shown in Fig. 9.11b [25–32]. By adjusting the standoff distance and design scheme of the wire mask, the fin array shape and density can be accurately controlled [25, 30–32]. The thermal and mechanical tests suggest that the CSAM fin arrays had superior

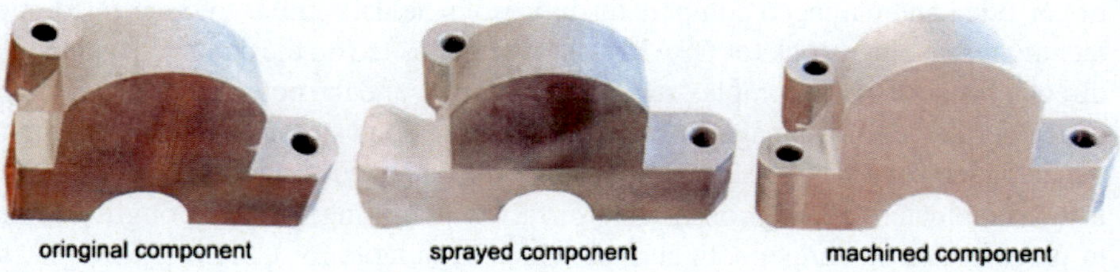

Fig. 9.10 Modification process of adding a new part to a bearing cap via CSAM [24]

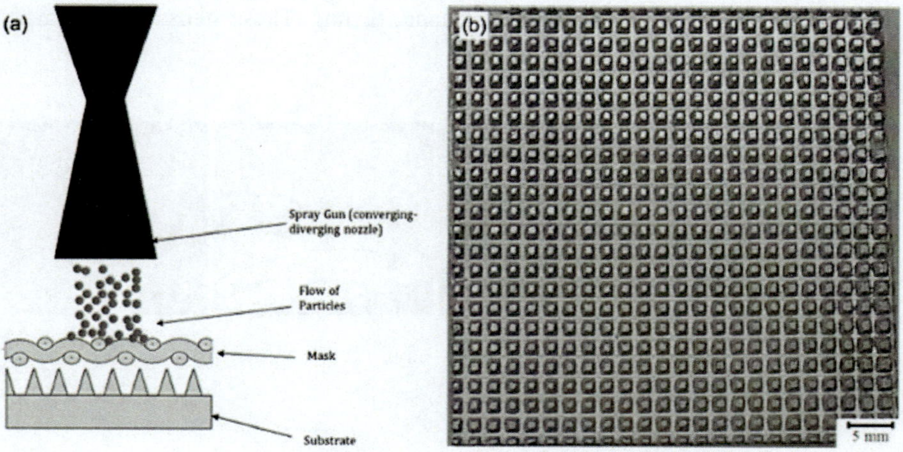

Fig. 9.11 CSAM pyramidal fin arrays heat sink. **a** schematic of the fabrication process and **b** digital photo of a CSAM fin array heat sink

thermal performance than traditional straight (rectangular) fins, due to the higher convective heat transfer coefficients [30, 31, 33]. Further studies have suggested that heat transfer performance can be improved by increasing the fin height, increasing fin density, or by using staggered configurations at the expense of a higher pressure loss [25, 28]. In addition, mask-assisted CSAM also can be used to fabricate chip heat sinks, metal markings, or other products with special patterns [34–36].

9.3 CSAM for Repairing Damaged Components

Component damage frequently occurs after long service times due to corrosion, wear, fatigue, or other causes. Damaged components often cannot be reclaimed or repaired, and have to be replaced due to the lack of effective restoration methods. CSAM is a cost-effective process that has shown great potential in the repair of damaged components, due to its capability to avoid any thermal damage to the underlying substrate material, and unique ability to retain the original properties of the feedstock powder. During the past decade, CSAM has been successfully applied to repair a variety of corroded and damaged components in various fields. Similar to other restoration techniques, the feedstock for CSAM cannot be deposited on the damaged component directly because of the complex surface topography, and the potential for the contaminated surface of the damaged area to compromise bond strength. Pre-machining on the damaged zone is necessary to reconstruct the damaged surface. Surface treatment is then conducted on the reconstructed surface with milling, grinding, or grit-blasting to prepare a clean and smooth surface which is suitable for CSAM [37–39]. After deposition, the as-coated components also have to be post-machined to their original dimensions. The standard restoration process normally includes the following four stages: pre-machining of the damaged zone, material deposition, post-machining on the back-filling material, and performance testing. These steps are illustrated in Fig. 9.12.

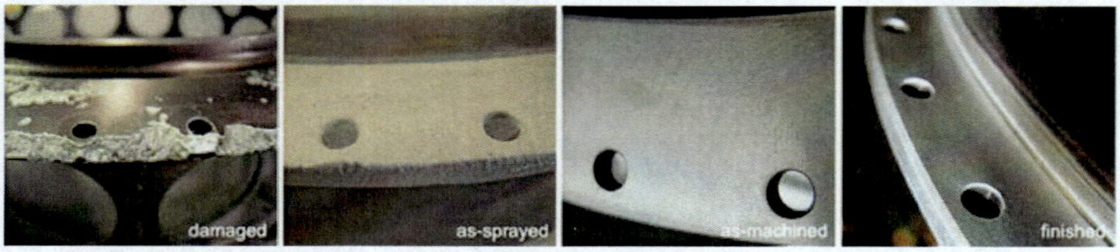

Fig. 9.12 Restoration procedure via CSAM

9.3.1 Corrosive and Erosive Damage Restoration

Among all applications of CSAM, the repair of aircraft components is potentially the most promising due to the severe corrosion and wear that aerospace components are subject to. This is due to the exposure of these components to high-speed impact, high-speed rotation and humid air. Magnesium and magnesium alloys possess many merits (e.g. high stiffness, low density, high thermal conductivity and excellent machinability) over other metals, and thus have been widely used for the fabrication of aircraft transmission gearboxes. However, as an electrochemically active material, magnesium components tend to corrode due to the anoidic reaction with other metallic materials. Therefore, magnesium gearboxes frequently suffer from electrochemical corrosion in service. Such corrosion significantly reduces the component lifetime, increasing the maintenance cost and potential failure risk. It is of great importance to properly repair the corroded zone and bring the restored component back to a serviceable condition. By using CSAM, the corroded part of the transmission gearbox can be repaired with aluminium or aluminium alloy materials. Figure 9.13 shows CSAM being used for repair of damaged transmission gearboxes [40–42]. Mechanical and corrosion tests on the repaired components confirm that the backfilled aluminium and aluminium alloy materials had excellent adhesive strength, wear-resistance, and corrosion-resistance. These restored components are suitable to be returned to flying [43–45].

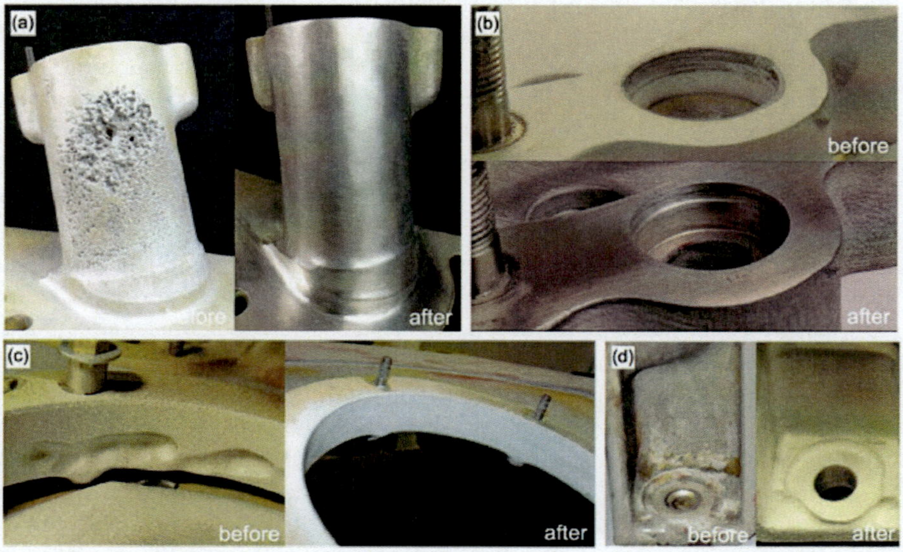

Fig. 9.13 Comparison between damaged and repaired components via CSAM: **a** S-92 helicopter gearbox sump, **b** oil tube bores in CH47 helicopter accessory cover, **c** UH-60 helicopter gearbox sump and **d** UH-60 rotor transmission housing [40–42]

Fig. 9.14 Comparison between damaged and repaired components via CSAM: **a** AH-64 helicopter mast support, **b** F18-AMAD gearbox and **c** front frame of T-700 engine [42, 46, 47]

Aluminium and aluminium alloys are also widely used in aircraft, marine and automotive components due to their low density, high tensile strength, good formability and excellent corrosion resistance. In the aerospace industry, the helicopter mast support is an example of a major component which is commonly manufactured using aluminium alloys. Corrosion pitting on the snap ring groove surface is a typical form of damage that helicopter mast supports can suffer from during their service life. When the pitting damage exceeds a criterion, the mast support must be repaired or replaced. Rather than scrapping the mast support, CSAM can be used to repair the damage and return the component to service. Figure 9.14a shows a comparison between a damaged and repaired snap ring groove surface of a mast support [46]. The detailed restoration procedure involves the blending out of corrosion pits, then backfilling the removed part by CSAM deposition of aluminium alloy, and finally post-machining the as-repaired component. Similar CSAM restoration work of aircraft aluminium alloy components is shown in Fig. 9.14b and c [42, 47]. In the marine industry, CSAM has been successfully applied to the repair of corroded internal bore surface of navy aluminium alloy valve actuator without thermally affecting the base material, showing improved capabilities when compared to other traditional repair methods (i.e., tungsten inert gas welding, metal inert gas welding, and laser cladding). Figure 9.15a shows a comparison between the damaged and repaired actuator. The actuator repaired using CSAM has passed all property tests and is now back in service [48]. Moreover, in the automotive industry, CSAM has been applied to repair corroded oil pump housings of Caterpillar-3116 and Caterpillar-3126 engines with aluminium alloy (Fig. 9.15b). It was reported that more than 30 corroded pump housings were restored via CSAM from 2012 to 2013 in Moscow. All of these reconditioned components have returned to service without any failure reports so far [49].

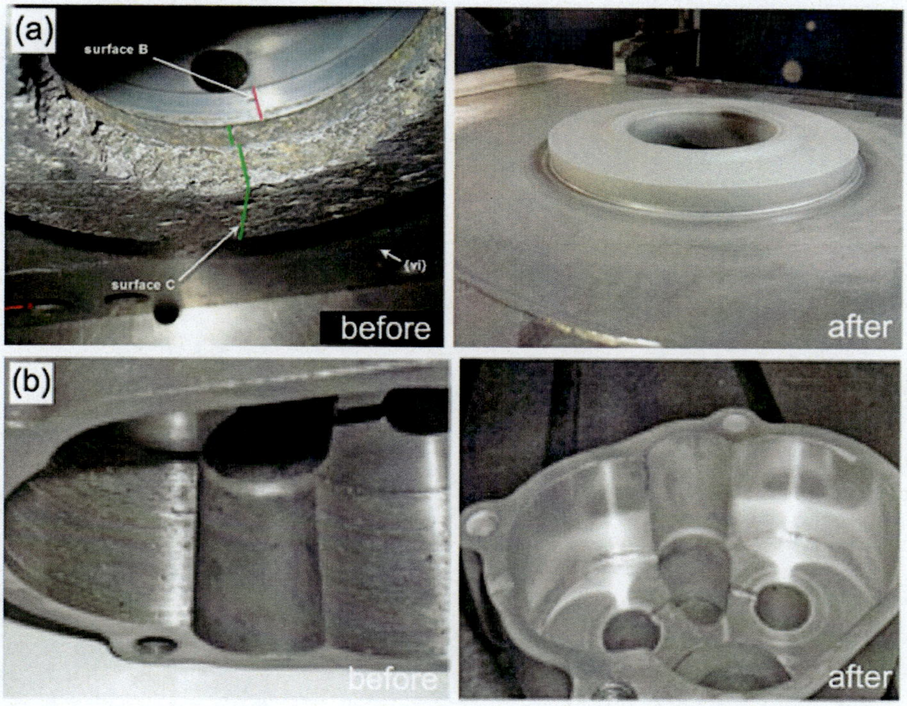

Fig. 9.15 Comparison between damaged and repaired components via CSAM: **a** internal bore surface of a navy valve actuator and **b** oil pump housing of Caterpillar engine [48, 49]

The nickel superalloy Inconel is another important material that is frequently used in aerospace industry, due to its excellent mechanical properties at high temperatures and resistance to corrosion. Aircraft components made using nickel alloys suffer from significant mechanical and thermal loading during flight. This results in severe wear and corrosion of these components, and thus reductions in their service lifetimes. Nose wheel steering actuator barrels manufactured using nickel alloy and are located in the front of the aircraft, which exposes these components to moisture and other harsh environments when the landing gear is extended. These negative factors affect the component joints, resulting in corrosion of landing gear. By using CSAM, a corroded B737 nose wheel steering actuator barrel was repaired with CSAM deposition of nickel alloy. Figure 9.16 shows the comparison between a damaged and repaired B737 nose wheel steering actuator barrel. The corroded surface becomes smooth after repair, without any pits and cracks [40].

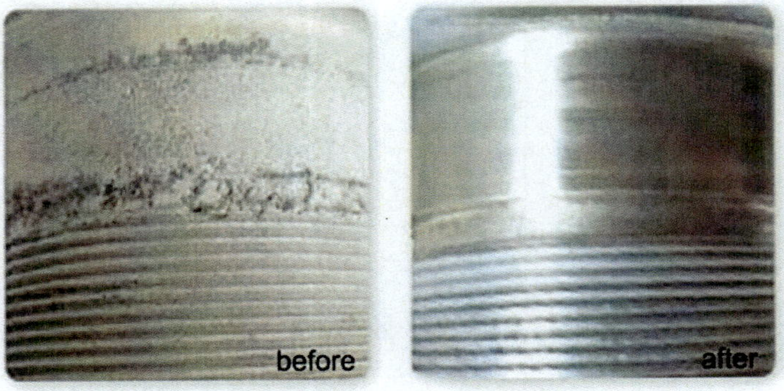

Fig. 9.16 Restored B737 nose wheel steering actuator barrel via CSAM [40]

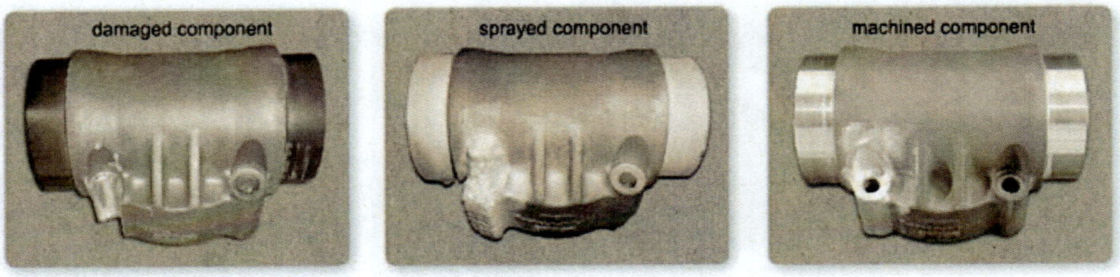

Fig. 9.17 Restoration process of a mechanically damaged flap transmission tee box housing of an aircraft via CSAM [40]

9.3.2 Mechanical Damage Restoration

Mechanical defects caused by fatigue, accidental impact, overhaul or improper operation are another type of damage that components may experience. This type of mechanical damage can also be repaired by CSAM. Figure 9.17 shows the restoration process of a mechanically damaged flap transmission tee box housing of an aircraft via CSAM [40]. The damaged component was repaired to the original condition it was in prior to the damage occurring. CSAM is also used to repair severe mechanical damage on large cast automotive parts, as can be seen in Fig. 9.18a [50]. Restoration of damaged molds is also in the scope of CSAM, as shown in Fig. 9.18b. The repaired mold presented similar results during the cutting force test to bulk material, and even better wear-resistance performance than the original component [51].

9.3.3 Damaged Metal Sheet Restoration

Commercial and military aircraft skins are primarily manufactured from aluminium alloy sheet, with additional aluminium cladding layer to prevent corrosion. During

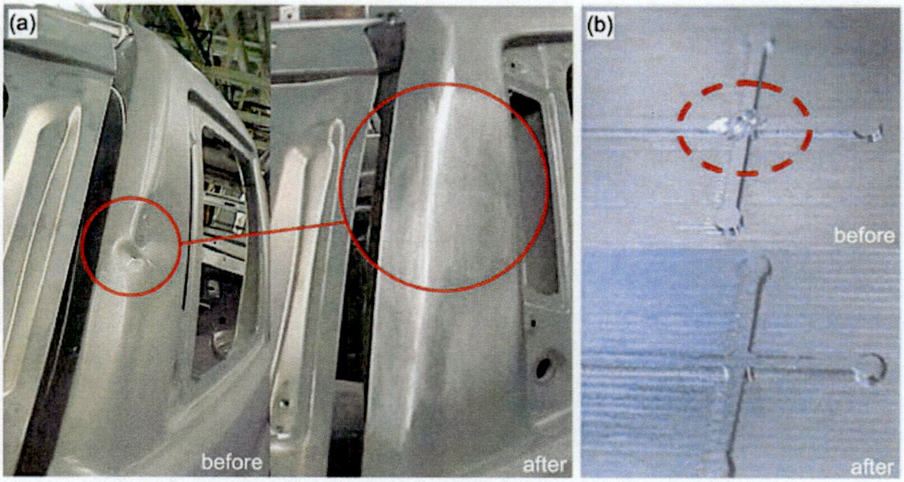

Fig. 9.18 Comparison between damaged and repaired components via CSAM: **a** a large cast automotive part and **b** a mold [50, 51]

routine flight, the aircraft skin has a high probability of suffering erosion and scratch damages due to high-velocity impact of dust and debris. The damaged area acts as a weak point, preferentially allowing the corrosion to penetrate through the protective aluminium clad layer, down to the base aluminium alloy. Once the base aluminium alloy is seriously corroded, the maintenance and failure cost will be very high. Therefore, it is necessary to repair the damaged aluminium cladding before the corrosion penetrates to the underlying base material. Some thermal spray technologies have been used to repair the damaged aircraft skin; however, the thermal sprayed coatings suffer from high porosity, high oxidation, and phase transformations during deposition. Most importantly, upon impact the high-temperature feedstock may have adversely effected the underlying base material, significantly degrading the aircraft skin performance [52]. CSAM has the potential to carry out this repair without damaging the underlying base material [53–55]. Figure 9.19 shows the repair of an aluminium cladding layer on an aluminium alloy panel via CSAM [56]. Visually, the damaged area is completely filled by the aluminium deposit without any obvious distinction from the surrounding aluminium cladding. Mechanical tests of the repaired sample suggest that the deposit hardness was higher than the original aluminium cladding, and the fatigue resistance of the repaired sample was improved. In addition, corrosion test reveals that the CSAM aluminium deposited on the damaged aluminium cladding provided an effective protection to the aluminium alloy base material, despite slight corrosion propagating from the repair edge in some cases [56].

In a recent study, fatigue tests of CSAM repaired aluminium alloy 2099 panels were conducted [57]. Figure 9.20a shows the digital photo of the machined notch on the aluminium alloy 2099 panel surface. The notch was repaired by CSAM with aluminium alloys 2198 and 7075, respectively. Crack growth tests were then

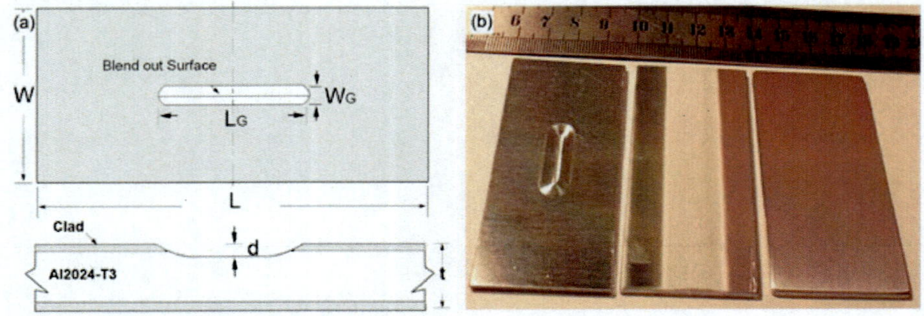

Fig. 9.19 Repair result of aluminium cladding layer on the aluminium alloy panel with self-machined damage via CSAM: **a** schematic of the self-machined damage and **b** restoration process [56]

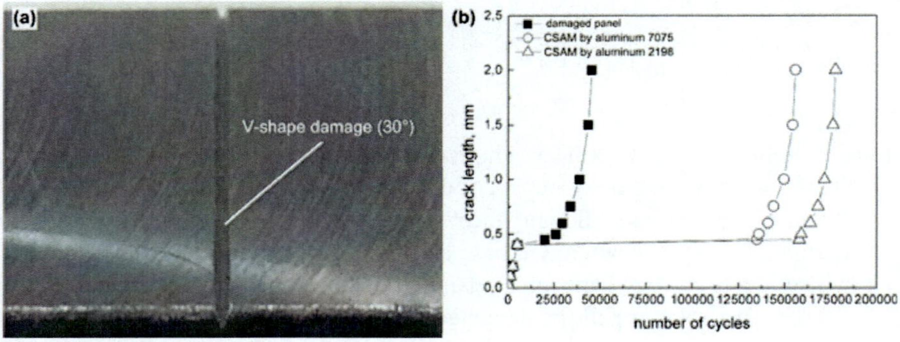

Fig. 9.20 **a** Digital photo of the machined notch made on the aluminium alloy 2099 panel surface, **b** crack length versus the number of loading cycles [57]

performed on these repaired panels [57]. Figure 9.20b shows the crack length as a function of the number of loading cycles for the damaged and repaired aluminium alloy panels. The crack length of the repaired panels was decreased by a factor of 6–7 when compared to the damaged panel. This is due to the strong adhesion between the CSAM deposit and the underlying substrate preventing crack initiation and growth. These results clearly demonstrate the feasibility of CSAM restoration in improving the fatigue performance of damaged sheets.

Aircraft fuselages also suffer from Multi Site Damage (MSD) issues. For nearly all kinds of aircraft, the fuselage is fabricated through lap joints of aluminium alloy skin panels. Therefore, failures and accidents associated with lap joints may occur after extended service lives. Fastener bore corrosion will take place through the interface between fastener and skin, despite the edge sealing between mating panels. In addition, multiple fastened strip repairs also act as a weak point, preferentially inducing the generation of cracks. These potential risks cumulatively cause the MSD on the aircraft skin. By using CSAM combined with standard sealant, the invasion of the moisture into the joint from the interface between mating panels and also between

Fig. 9.21 Cold spray deposit over the simulated fuselage fasteners

fasteners and aircraft skin can be prevented [53, 54, 58]. Figure 9.21 shows a digital photo of CSAM deposit over the simulated fuselage fasteners. It is shown that the surface of the fasteners are completely sealed by the CSAM deposit. Subsequent fatigue testing results indicate that the deposit remained intact even when the underlying fuselage joint skin experienced MSD. Furthermore, CSAM has shown great potential in the repair of cracks caused by MSD, which helps to enhance the fuselage structural integrity [53–55].

Aircraft propeller blades also suffer from serious erosion due to high-velocity impact by dust, debris and water droplets in service. Moreover, relatively large debris impacts during aircraft taking off and landing will lead to more severe damage to the blade surface. The current method used to repair this damage is simply to grind the eroded blade to a level that the damage is removed. By using CSAM, the material removed during the grinding process can be filled, which allows for the original blade dimensions to be restored. Figure 9.22 illustrates the repair process of aluminium alloy blades via CSAM. The restored blades have passed an extensive airworthiness test program and are currently back in service [59].

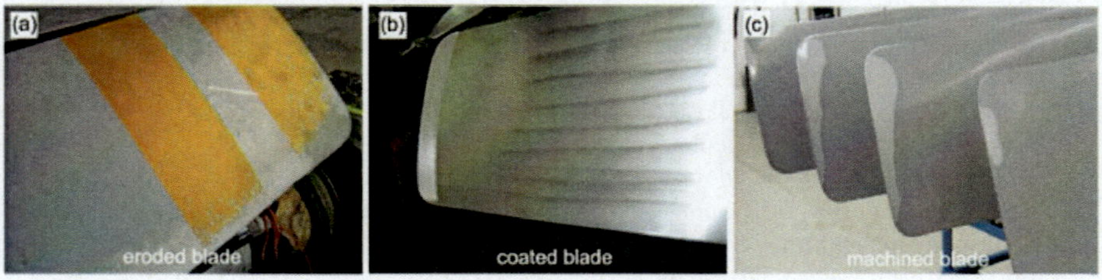

Fig. 9.22 Restoration process of aluminium alloy blades via CSAM [59]

References

1. Yin, S., Wang, X., Li, W., Liao, H., Jie, H.: Deformation behavior of the oxide film on the surface of cold sprayed powder particle. Appl. Surf. Sci. **259**, 294–300 (2012). https://doi.org/10.1016/j.apsusc.2012.07.036
2. Czampa, M., Markos, S., Szalay, T.: Improvement of drilling possibilities for machining powder metallurgy materials. Procedia CIRP. **7**, 288–293 (2013). https://doi.org/10.1016/j.procir.2013.05.049
3. Bordin, A., Ghiotti, A., Bruschi, S., Facchini, L., Bucciotti, F.: Machinability characteristics of wrought and EBM CoCrMo alloys. Procedia CIRP. **14**, 89–94 (2014). https://doi.org/10.1016/j.procir.2014.03.082
4. Aziz, M.S.A., Ueda, T., Furumoto, T., Abe, S., Hosokawa, A., Yassin, A.: Study on machinability of laser sintered materials fabricated by layered manufacturing system: Influence of different hardness of sintered materials. Procedia CIRP. **4**, 79–83 (2012). https://doi.org/10.1016/j.procir.2012.10.015
5. Montevecchi, F., Grossi, N., Takagi, H., Scippa, A., Sasahara, H., Campatelli, G.: Cutting forces analysis in additive manufactured AISI H13 alloy. Procedia CIRP. **46**, 476–479 (2016). https://doi.org/10.1016/j.procir.2016.04.034
6. Aldwell, B., Kelly, E., Wall, R., Amaldi, A., O'Donnell, G.E., Lupoi, R.: Machinability of Al 6061 deposited with cold spray additive manufacturing. J. Therm. Spray Technol. **26**, 1573–1584 (2017). https://doi.org/10.1007/s11666-017-0586-x
7. Tutunea-Fatan, O.R., Fakhri, M.A., Bordatchev, E.V.: Porosity and cutting forces: from macroscale to microscale machining correlations. Proc. Inst. Mech. Eng. Part B J. Eng. Manuf. **225**, 619–630 (2011). https://doi.org/10.1177/2041297510394057
8. Abolghasemi Fakhri, M., Bordatchev, E.V., Tutunea-Fatan, O.R.: An image-based methodology to establish correlations between porosity and cutting force in micromilling of porous titanium foams. Int. J. Adv. Manuf. Technol. **60**, 841–851 (2012). https://doi.org/10.1007/s00170-011-3647-1
9. Lu, G., Lu, G.Q., Xiao, Z.M.: Mechanical properties of porous materials. J. Porous Mater. **6**, 359–368 (1999). https://doi.org/10.1023/A:1009669730778
10. Ramulu, M., Rao, P.N., Kao, H.: Drilling of (Al2O3)p/6061 metal matrix composites. J. Mater. Process. Technol. **124**, 244–254 (2002). https://doi.org/10.1016/S0924-0136(02)00176-0
11. Sova, A., Courbon, C., Valiorgue, F., Rech, J., Bertrand, P.: Effect of turning and ball burnishing on the microstructure and residual stress distribution in stainless steel cold spray deposits. J. Therm .Spray Technol. **26**, 1922–1934 (2017). https://doi.org/10.1007/s11666-017-0655-1
12. MacDonald, D., Fernández, R., Delloro, F., Jodoin, B.: Cold spraying of armstrong process titanium powder for additive manufacturing. J. Therm. Spray Technol. (2016). https://doi.org/10.1007/s11666-016-0489-2
13. Abreeza, M., Yuji, I., Kazuhiro, O.: Computational Simulation for Cold Sprayed Deposition. Sendai, Japan (2011)
14. Barnett, B., Trexler, M., Champagne, V.: Cold sprayed refractory metals for chrome reduction in gun barrel liners. Int. J. Refract. Met. Hard Mater. **53**, 139–143 (2015). https://doi.org/10.1016/j.ijrmhm.2015.07.007
15. Choi, H.J., Lee, M., Lee, J.Y.: Application of a cold spray technique to the fabrication of a copper canister for the geological disposal of CANDU spent fuels. Nucl. Eng. Des. **240**, 2714–2720 (2010). https://doi.org/10.1016/j.nucengdes.2010.06.038
16. Sova, A., Grigoriev, S., Okunkova, A., Smurov, I.: Potential of cold gas dynamic spray as additive manufacturing technology. Int. J. Adv. Manuf. Technol. **69**, 2269–2278 (2013). https://doi.org/10.1007/s00170-013-5166-8
17. Kumar, S., Chavan, N.M.: Cold Spray Coating Technology : Activities at ARCI. Hyderabad, India (2011)
18. Chad, H.: CSIRO Titanium Technologies and Additive Manufacturing (2014)
19. Richter, P.: New value chain for advanced coatings by using cold spray. Greece, Athens (2014)

20. May, C., Marx, S., Paul, A.: Selected R&D Results and Industrial Applications. Presented at the (2013)
21. Sova, A., Okunkova, A., Grigoriev, S., Smurov, I.: Velocity of the particles accelerated by a cold spray micronozzle: experimental measurements and numerical simulation. J. Therm. Spray Technol. **22**, 75–80 (2013). https://doi.org/10.1007/s11666-012-9846-y
22. Briefing, T.: Metal coated particles and defense applications. In: CSAT Workshop. Worcester, USA (2013)
23. Lynch, M.E., Gu, W., El-Wardany, T., Hsu, A., Viens, D., Nardi, A., Klecka, M.: Design and topology/shape structural optimisation for additively manufactured cold sprayed components. Virtual Phys. Prototyp. **8**, 213–231 (2013). https://doi.org/10.1080/17452759.2013.837629
24. Villafuerte, J.: Using Cold Spray to Add Features to Components (2015)
25. Cormier, Y., Dupuis, P., Farjam, A., Corbeil, A., Jodoin, B.: Additive manufacturing of pyramidal pin fins: height and fin density effects under forced convection. Int. J. Heat Mass Transf. **75**, 235–244 (2014). https://doi.org/10.1016/j.ijheatmasstransfer.2014.03.053
26. Cormier, Y., Dupuis, P., Jodoin, B., Corbeil, A.: Pyramidal fin arrays performance using streamwise anisotropic materials by cold spray additive manufacturing. J. Therm. Spray Technol. **25**, 170–182 (2016). https://doi.org/10.1007/s11666-015-0267-6
27. Dupuis, P., Cormier, Y., Fenech, M., Corbeil, A., Jodoin, B.: Flow structure identification and analysis in fin arrays produced by cold spray additive manufacturing. Int. J. Heat Mass Transf. **93**, 301–313 (2016). https://doi.org/10.1016/j.ijheatmasstransfer.2015.10.019
28. Dupuis, P., Cormier, Y., Fenech, M., Jodoin, B.: Heat transfer and flow structure characterization for pin fins produced by cold spray additive manufacturing. Int. J. Heat Mass Transf. **98**, 650–661 (2016). https://doi.org/10.1016/j.ijheatmasstransfer.2016.03.069
29. Cormier, Y., Dupuis, P., Jodoin, B., Ghaei, A.: Finite element analysis and failure mode characterization of pyramidal fin arrays produced by masked cold gas dynamic spray. J. Therm. Spray Technol. **24**, 1549–1565 (2015). https://doi.org/10.1007/s11666-015-0317-0
30. Farjam, A., Cormier, Y., Dupuis, P., Jodoin, B., Corbeil, A.: Influence of alumina addition to aluminum fins for compact heat exchangers produced by cold spray additive manufacturing. J. Therm. Spray Technol. **24**, 1256–1268 (2015). https://doi.org/10.1007/s11666-015-0305-4
31. Cormier, Y., Dupuis, P., Jodoin, B., Corbeil, A.: Net shape fins for compact heat exchanger produced by cold spray. J. Therm. Spray Technol. **22**, 1210–1221 (2013). https://doi.org/10.1007/s11666-013-9968-x
32. Cormier, Y., Dupuis, P., Jodoin, B., Corbeil, A.: Mechanical properties of cold gas dynamic-sprayed near-net-shaped fin arrays. J. Therm. Spray Technol. **24**, 476–488 (2014). https://doi.org/10.1007/s11666-014-0203-1
33. Kotoban, D., Grigoriev, S., Okunkova, A., Sova, A.: Influence of a shape of single track on deposition efficiency of 316L stainless steel powder in cold spray. Surf. Coatings Technol. **309**, 2–9 (2016). https://doi.org/10.1016/j.surfcoat.2016.10.052
34. Birtch, W.: Practical Cold Spray (2010)
35. Bierk, B., Mgr, G., Addison, G., Elmquist, B.: Repair technology development projects. In: CSAT Workshop. Worcester, USA (2011)
36. Kashirin, A., Klyuev, O., Buzdygar, T., Shkodkin, A.: Modern applications of the low pressure cold spray. In: Proceedings of the 2011 International Thermal Spray Conference. Hamburg (2011)
37. Yin, S., Xie, Y., Suo, X., Liao, H., Wang, X.: Interfacial bonding features of Ni coating on Al substrate with different surface pretreatments in cold spray. Mater. Lett. **138**, 143–147 (2015). https://doi.org/10.1016/j.matlet.2014.10.016
38. Nastic, A., Vijay, M., Tieu, A., Rahmati, S., Jodoin, B.: Experimental and numerical study of the influence of substrate surface preparation on adhesion mechanisms of aluminum cold spray coatings on 300M steel substrates. J. Therm. Spray Technol. **26**, 1461–1483 (2017). https://doi.org/10.1007/s11666-017-0602-1
39. Sun, W., Tan, A.W.Y., Khun, N.W., Marinescu, I., Liu, E.: Effect of substrate surface condition on fatigue behavior of cold sprayed Ti6Al4V coatings. Surf. Coatings Technol. **320**, 452–457 (2017). https://doi.org/10.1016/j.surfcoat.2016.11.093

40. Schell, J.: Cold spray aerospace applications. In: CSAT Workshop. X, Worcester, USA (2016)
41. Howe, C.: Cold spray repair of the CH-47 accessory cover. In: CSAT Workshop. Worcester, USA (2014)
42. Kilchenstein, G.: Cold spray technologies used for repair. In: JTEG Monthly Teleconference (2014)
43. DeForce, B., Eden, T., Potter, J., Champagne, V., Leyman, P., Helfritch, D.: Application of aluminum coatings for the corrosion protection of magnesium by cold spray. In: Tri-Service Corrosion Conference. Denver, USA (2007)
44. Champagne, V.K., Helfritch, D., Leyman, P.F.: Magnesium repair by cold spray. Plat. Surf. Finish. **95**, 19–28 (2008)
45. Security, E., Program, T.C.: Cost and Performance Report Cold Spray for Repair of Magnesium (2011)
46. Leyman, P.F., Champagne, V.K.: Cold Spray Process Development for the Reclamation of the Apache Helicopter Mast Support (2009)
47. Howe, C.: Cold spray qualification of T700 engine front frame. In: CSAT Workshop. Worcester, USA (2015)
48. Widener, C.A., Carter, M.J., Ozdemir, O.C., Hrabe, R.H., Hoiland, B., Stamey, T.E., Champagne, V.K., Eden, T.J.: Application of high-pressure cold spray for an internal bore repair of a navy valve actuator. J. Therm. Spray Technol. **25**, 193–201 (2016). https://doi.org/10.1007/s11666-015-0366-4
49. Lyalyakin, V.P., Kostukov, A.Y., Denisov, V.A.: Special features of reconditioning the housing of a Caterpillar diesel oil pump by gas-dynamic spraying. Weld. Int. **30**, 68–70 (2016). https://doi.org/10.1080/09507116.2015.1030152
50. Maev, R.G., Strumban, E., Leshchinskiy, V., Dzhurinskiy, D.: Repair applications of the LPCS process. In: CSAT Workshop. Worcester, USA (2014)
51. Lee, J.C., Kang, H.J., Chu, W.S., Ahn, S.H.: Repair of damaged mold surface by cold-spray method. CIRP Ann. - Manuf. Technol. **56**, 577–580 (2007). https://doi.org/10.1016/j.cirp.2007.05.138
52. D6–51343, B.S.: Thermal Spray Repair of Exterior Clad Aluminum. USA (2006)
53. Jones, R., Molent, L., Barter, S., Matthews, N., Tamboli, D.: Supersonic particle deposition as a means for enhancing the structural integrity of aircraft structures. Int. J. Fatigue. **68**, 260–268 (2014). https://doi.org/10.1016/j.ijfatigue.2014.03.013
54. Matthews, N., Jones, R., Sih, G.C.: Application of supersonic particle deposition to enhance the structural integrity of aircraft structures. Sci. China Phys. Mech. Astron. **57**, 12–18 (2014). https://doi.org/10.1007/s11433-013-5367-z
55. Jones, R., Matthews, N., Rodopoulos, C.A., Cairns, K., Pitt, S.: On the use of supersonic particle deposition to restore the structural integrity of damaged aircraft structures. Int. J. Fatigue. **33**, 1257–1267 (2011). https://doi.org/10.1016/j.ijfatigue.2011.03.013
56. Yandouzi, M., Gaydos, S., Guo, D., Ghelichi, R., Jodoin, B.: Aircraft skin restoration and evaluation. J. Therm. Spray Technol. **23**, 1281–1290 (2014). https://doi.org/10.1007/s11666-014-0130-1
57. Cavaliere, P., Silvello, A.: Crack repair in aerospace aluminum alloy panels by cold spray. J. Therm. Spray Technol. **26**, 661–670 (2017). https://doi.org/10.1007/s11666-017-0534-9
58. Jones, R., Matthews, N., Elston, J., Cairns, K., Baker, J., Wadsley, B., Pitt, S.: Spd repairs to thin aluminium structures. In: 28th International Congress of the Aeronautical Sciences. pp. 1–9 (2012)
59. Stoltenhoff, T., Zimmermann, F., Surface, P., Gmbh, T.: LOXPlate ® Coatings for Aluminum Aerospace Components Exposed to High Dynamic Stresses. Ratingen, Germany (2012)

Chapter 10
Summary and Outlooks

Abstract CSAM as an emerging additive manufacturing process has been given special attentions by researchers and engineers worldwide. This new additive manufacturing process has shown great potentials in fabricating free-stranding components and repairing damage components. To date, a large number of studies have been carried out to explore the bonding mechanism, microstructures and properties of cold sprayed deposits. Various methods have been developed to strengthen the cold sprayed deposits so that they can be used by industry. This chapter summarizes the main findings of this book, and the future perspectives of CSAM are also proposed.

10.1 Summary

Cold spray deposition has been developed as an additive manufacturing technology only in recent years. As a new additive manufacturing process, CSAM retains all the benefits of cold spray technology. As compared with fusion-based additive manufacturing process such as LBM, EBM and LMD, the formation of a cold sprayed deposit relies mainly on the particle kinetic energy prior to impact rather than the thermal energy. The feedstock used for cold spray remains solid state during the entire deposition process. Deposition is achieved through local metallurgical bonding and mechanical interlocking which are caused by localized plastic deformation at the inter-particle and particle–substrate interfaces. This consolidation mechanism renders CSAM a unique microstructure that is free of defects commonly encountered in high-temperature processes, such as high residual thermal stress and phase transformation. In addition, CSAM offers shorter production times, unlimited product size, and high flexibility. Currently, CSAM has been primarily used for fabricating components with simple geometry (e.g., walls, tubes and flanges). Components with complex structures can also be fabricated via CSAM, but needs further post-machining process as a finish. CSAM is also widely used to repair damaged components due to its high flexibility, lack of adverse effects on the underlying substrate material, and to retain the original properties of the feedstock. Overall, CSAM is a promising technology that can be used in a broad range of industrial sectors in the future. This book aims to provide a comprehensive introduction of CSAM from its

S. Yin and R. Lupoi, *Cold Spray Additive Manufacturing*, Springer Tracts in Additive Manufacturing, https://doi.org/10.1007/978-3-030-73367-4_10

fundamentals to applications. The main content and contribution of this book are summarized as follows.

In Chap. 1, the working principles of CSAM, commercial cold spray system, comparison between CSAM and other additive manufacturing processes and brief introduction of the applications of CSAM are presented.

Chapter 2 provides a comprehensive introduction on the gas flow, particle acceleration, and heat transfer behavior in cold spray. In CSAM, deposition is achieved through the high-velocity impact of powder particles onto substrate, and the acceleration of the particles is released through a convergent-divergent nozzle. Therefore, a thorough understanding of particle acceleration and heating behavior inside and outside a cold spray nozzle is critical for developing high-performance cold sprayed deposits. This chapter summarizes and reviews the most important scientific work with regard to experimental measurements and numerical simulations in the topic of aerodynamics that are involved in cold spray.

In Chap. 3, key manufacturing parameters that are critical for the operation of CSAM are introduced, including gas pressure, gas temperature, gas type, powder feed rate, nozzle traverse speed, scanning step, standoff distance, spray angle and trajectory. Their roles in cold spray deposition are introduced and the relationship among these parameters are discussed. These parameters must work in unison and under careful control to produce high-quality deposits with desirable microstructure and excellent performance.

In Chap. 4, the microstructures of cold sprayed deposits are interpreted and discussed. Cold sprayed deposits typically have heterogenous microstructure and grain structures. A large number of dislocations and newly formed grains due to the severe plastic deformation are localized in the shear zones at the interparticle interfaces, while in the interior region the microstructure is mainly characterized by coarse grains. This chapter thoroughly discusses the formation mechanism of dislocations and UFGs in various cold sprayed deposits during cold spray deposition. The formation mechanisms of intermetallic compounds and amorphous phases are also discussed.

Chapter 5 provides an introduction of commonly used strengthening strategies for improving the mechanical properties of cold sprayed deposits. Cold sprayed deposits normally have unfavorable mechanical properties in their as-fabricated state as compared to conventional fusion-based additive manufactured counterparts due to the inherent defects (i.e., porosity and unbonded interparticle boundary) in the cold sprayed deposits. As structural weaknesses, theses defects will impair the mechanical properties of cold sprayed deposit. In this chapter, various strengthening methods including conventional annealing, hot isostatic pressing, hot rolling, in-situ shot peening, in-situ densification, in-situ laser assistance, friction stirring and powder heat treatment are introduced and their strengthening mechanisms are discussed.

Chapters 6, 7 and 8 aim to introduce the application of CSAM for fabricating special materials including metal matrix composites (MMCs), nanocrystalline metals, amorphous metals and high entropy alloys (HEAs). These materials are either composite materials or non-conventional metals and alloys; they are mostly difficult to be deposited via cold spray due to the nature of the starting powder feedstock. For example, the feedstock for fabricating MMC deposits must be a mixture of two or more dissimilar materials, and it is hard to accurately control the ratio between different components in the deposit. For nanocrystalline metals, amorphous metals and HEAs, the starting powders typically have high hardness and strength, which significantly reduces their cold sprayability and potentially leads to the deposit having a large number of defects. These chapters comprehensively explain how these unconventional metallic materials are fabricated via cold spray and their microstructure and properties after cold spray.

In Chapter 9, the current and potential applications of CSAM in industry are presented. CSAM can be used in a broad range of industrial sectors such as aerospace, aviation, marine, automotive and nuclear. It is capable of fabricating free-stranding components and repairing damage components. Cold sprayed deposits typically have a rough, undulating and porous surface and low dimensional accuracy, and thus they are not suitable for immediate use. This necessitates the use of a machining or finishing process. Therefore, this chapter aims to provide a thorough discussion on the post machining and the industrial applications of cold sprayed deposits.

10.2 Outlooks

It has been approximately 25 years since the first invention of cold spray. The method has triggered a wide acceptance at academic and industrial level with the development of a large number of research projects that have technically and effectively brought cold spray to where it is now. This book has described in detail one of the most interesting aspect, that is its application as additive manufacturing (CSAM). The field of additive manufacturing is growing in relevance and market shares in comparison to other processes, and there is no doubt that the book has demonstrated that CSAM will play a key role in this. CSAM is not intended to be, by large, a competitor to other metallic additive processes such as SLM and LBM. This book has clearly highlighted the process capabilities and advantages, for example the very fast build rate and the melt-free or limited melting nature of the particle bonding mechanism. Thus, there is and will be applications for CSAM is areas such as functionally-graded components, repair of oxide sensitive materials, multi-material parts that are not metallurgically compatible and hybrid manufacturing. The fact that the process is characterized by high build rates, can compensate for the fact of being in general more expensive versus other methods in terms of total cost per unit time. However, there are still disadvantages that remains unsolved. An example is the overall geometrical inaccuracy, leaving plenty of open research avenues in order to improve the process. We do therefore expect CSAM to grow in relevance, especially when combined with a post-process option such an heat-treatment (typically with pressure or in vacuum) that is aimed at homogenizing the part properties and return a level of elongation before fracture that is comparable to the bulk. Another example is advanced modeling technique which can allow the simulation of microstructure evolution during cold spray particle deposition. These, combined with the equipment development that is running in parallel and led by companies and institutions across the globe, will provide the necessary ingredients for an interesting success story in the years to come.